Information and Instructi

This shop manual contains several sections each covering a specific group of wheel type tractors. The Tab Index on the preceding page can be used to locate the section pertaining to each group of tractors. Each section contains the necessary specifications and the brief but terse procedural data needed by a mechanic when repairing a tractor on which he has had no previous actual experience.

Within each section, the material is arranged in a systematic order beginning with an index which is followed immediately by a Table of Condensed Service Specifications. These specifications include dimensions, fits, clearances and timing instructions. Next in order of arrangement is the procedures paragraphs.

In the procedures paragraphs, the order of presentation starts with the front axle system and steering and proceeding toward the rear axle. The last paragraphs are devoted to the power take-off and power lift sys-

tems. Interspersed w.   ar specifications pertaining to wear limits, torquing, etc.

HOW TO USE THE INDEX

Suppose you want to know the procedure for R&R (remove and reinstall) of the engine camshaft. Your first step is to look in the index under the main heading of ENGINE until you find the entry "Camshaft." Now read to the right where under the column covering the tractor you are repairing, you will find a number which indicates the beginning paragraph pertaining to the camshaft. To locate this wanted paragraph in the manual, turn the pages until the running index appearing on the top outside corner of each page contains the number you are seeking. In this paragraph you will find the information concerning the removal of the camshaft.

More information available at haynes.com
Phone: 805-498-6703

J H Haynes & Co. Ltd.
Haynes North America, Inc.

ISBN-10: 0-87288-079-6
ISBN-13: 978-0-87288-079-5

SHOP MANUAL
JOHN DEERE

SERIES
1020 1520 1530 2020 2030

Tractor serial number is located on right side of transmission case. Engine serial number is stamped on a plate at lower right front of engine cylinder block.

INDEX (By Starting Paragraph)

CONDENSED SERVICE DATA

	1020 Gasoline	1020 Diesel	1520 Gasoline	1520 Diesel	1530 Diesel
GENERAL					
Engine Make	Own	Own	Own	Own	Own
Number Cylinders	3	3	3	3	3
Bore—Inches	3.86	3.86	4.02	4.02	4.02
Stroke—Inches	3.86	4.33	4.33	4.33	4.33
Displacement—Cubic Inches	136	152	164	164	164
Compression Ratio	7.5:1	16.3:1	8.0:1	16.3:1	16.2:1
Battery Terminal Grounded	Neg.	Neg.	Neg.	Neg.	Neg.
Forward Speeds	8	8	8	8	8
TUNE-UP					
Firing order	1-2-3	1-2-3	1-2-3	1-2-3	1-2-3
Compression Pressure—PSI @ 200 RPM	120	300	120	300	300
Valve Clearance:					
Inlet—	0.014	0.014	0.014	0.014	0.014
Exhaust—	0.022	0.018	0.022	0.018	0.018
Timing Mark Location	Crankshaft Pulley	Crankshaft Pulley	Crankshaft Pulley	Crankshaft Pulley	. . . .
Breaker Point Gap	0.020	. . . .	0.020	. . . .	. . . .
Spark Plug Size	14MM	. . . .	14MM	. . . .	. . . .
Electrode Gap	0.025	. . . .	0.025	. . . .	. . . .
Engine Low Idle—RPM	600	800	600	800	650
Engine High Idle—RPM	2680	2650	2680	2650	2650
Working Range	1500-2500	1500-2500	1500-2500	1500-2500	1500-2500
PTO Horsepower @ 2500 RPM	38.8	38.9	47.9	46.5	46
SIZES—CAPACITIES— CLEARANCES					
Crankshaft Journal Diameter	3.1235-3.1245	3.1235-3.1245	3.1235-3.1245	3.1235-3.1245	3.1235-3.1245
Crankpin Diameter	2.309	2.7485	2.7485	2.7485	2.7485
Balancer Shaft Journal Diameter	. . . .	. . . .	. . . .	. . . .	. . . .
Piston Pin Diameter	1.1877	1.1877	1.1877	1.1877	See Paragraph 54
Main Bearing Clearance	0.001-0.0041	0.001-0.0041	0.001-0.0041	0.001-0.0041	0.001-0.0041
Rod Bearing Clearance	0.0012-0.0044	0.0012-0.0044	0.0012-0.0044	0.0012-0.0044	0.0012-0.0044
Camshaft Journal Clearance	0.0035-0.0055	0.0035-0.0055	0.0035-0.0055	0.0035-0.0055	0.0035-0.0055
Balancer Shaft Bearing Clearance	. . . .	. . . .	. . . .	. . . .	. . . .
Crankshaft End Play	0.002-0.008	0.002-0.008	0.002-0.008	0.002-0.008	0.002-0.008
Camshaft End Play	0.0025-0.0085	0.0025-0.0085	0.0025-0.0085	0.0025-0.0085	0.0025-0.0085
Piston Skirt Clearance	See Paragraph 53	See Paragraph 53	See Paragraph 53	See Paragraph 53	See Paragraph 53
Cooling System—Qts.	11	11	12	12	10.5
Crankcase (with filter)—Qts.	6	6	6	6	6
Fuel Tank—Gallons	16½	16½	19½	19½	19½
Trans. & Hydraulic Systems—Gals.	10	10	10	10	9½
TIGHTENING TORQUES— FT.-LBS.					
Cylinder Head	110	110	110	110	110
Main Bearings	85	85	85	85	110
Con. Rod Bearings (Oiled)	45	70	45	70	65
Rocker Arm Assembly	35	35	35	35	35
Flywheel	85	85	85	85	85

CONDENSED SERVICE DATA

	2020 Gasoline	2020 Diesel	2030 Gasoline	2030 Diesel
GENERAL				
Engine Make	Own	Own	Own	Own
Number Cylinders	4	4	4	4
Bore—Inches	3.86	3.86	4.02	4.02
Stroke—Inches	3.86	4.33	4.33	4.33
Displacement—Cubic Inches	180	202	219	219
Compression Ratio	7.5:1	16.3:1	7.5:1	16.3:1
Battery Terminal Grounded	Neg.	Neg.	Neg.	Neg.
Forward Speeds	8	8	8	8
TUNE-UP				
Firing order	1-3-4-2	1-3-4-2	1-3-4-2	1-3-4-2
Compression Pressure—PSI @ 200 RPM	120	300	120	300
Valve Clearance:				
Inlet—	0.014	0.014	0.014	0.014
Exhaust—	0.022	0.018	0.022	0.018
Timing Mark Location	Crankshaft Pulley	Crankshaft Pulley	Crankshaft Pulley	Crankshaft Pulley
Breaker Point Gap	0.020	....	0.020	....
Spark Plug Size	14MM	....	14MM	....
Electrode Gap	0.025	....	0.025	....
Engine Low Idle—RPM	600	800	600	800
Engine High Idle—RPM	2680	2650	2680	2650
Working Range	1500-2500	1500-2500	1500-2500	1500-2500
PTO Horsepower @ 2500 RPM	53.9	54.1	60.34	60.65
SIZES—CAPACITIES—CLEARANCES				
Crankshaft Journal Diameter	3.1235-3.1245	3.1235-3.1245	3.1230-3.1240	3.1230-3.1240
Crankpin Diameter	2.309	2.7485	2.3095	2.7485
Balancer Shaft Journal Diameter	1.500	1.500	1.500	1.500
Piston Pin Diameter	See Paragraph 54	1.1877	1.1877	See Paragraph 54
Main Bearing Clearance	0.001-0.0041	0.001-0.0041	0.0016-0.0046	0.0016-0.0046
Rod Bearing Clearance	0.0012-0.0044	0.0012-0.0044	0.0012-0.0044	0.0012-0.0044
Camshaft Journal Clearance	0.0035-0.0055	0.0035-0.0055	0.0035-0.005	0.0035-0.005
Balancer Shaft Bearing Clearance	0.0015-0.0045	0.0015-0.0045	0.0015-0.0045	0.0015-0.0045
Crankshaft End Play	0.002-0.008	0.002-0.008	0.002-0.008	0.002-0.008
Camshaft End Play	0.0025-0.0085	0.0025-0.0085	0.0025-0.0085	0.0025-0.0085
Piston Skirt Clearance	See Paragraph 53	See Paragraph 53	See Paragraph 53	See Paragraph 53
Cooling System—Qts.	12	12	12	12
Crankcase (with filter)—Qts.	6	6	6	6
Fuel Tank—Gallons	19½	19½	19½	19½
Trans. & Hydraulic Systems—Gals.	10	10	10	10
TIGHTENING TORQUES—Ft.-LBS.				
Cylinder Head	110	110	110	110
Main Bearings	85	85	85	85
Con. Rod Bearings (Oiled)	45	70	65	65
Rocker Arm Assembly	35	35	35	35
Flywheel	85	85	85	85

FRONT SYSTEM

AXLE AND SUPPORT

All Models

1. **AXLE CENTER MEMBER.** Center axle unit (5 or 5A—Fig. 1) attaches to front support by pivot bolt (6) and rear pivot pin (7). End clearance of pivot is controlled by shims (3). Five 0.015 shims are installed at factory assembly and recommended maximum end play is 0.015. Removing shims reduces the end play. Thrust washers (2) at front and rear of pivot bushing (4) are identical and may be interchanged to compensate for wear when unit is removed. Rear pivot pin (7) is pressed into front support (1—Fig. 2).

Steering bellcrank (13—Fig. 1) should have 0.001-0.006 clearance in the two bushings (12) pressed into center axle unit. Recommended end play of 0.010 is controlled by shims (11) which are 0.010 in thickness. Two

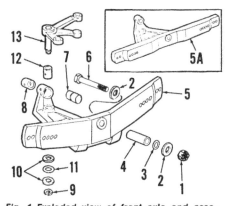

Fig. 1–Exploded view of front axle and associated parts. Series 2030 tractors may also be equipped with a one-piece fixed tread axle which is similar in major details.

1. Nut
2. Washer
3. Shim
4. Bushing
5. Axle
5A. Axle
6. Pivot bolt
7. Pivot pin
8. Bushing
9. Snap ring
10. Washer
11. Shim
12. Bushing
13. Steering bellcrank

shims are normally used; install additional shims when necessary providing snap ring (9) can still be installed.

2. **SPINDLES AND BUSHINGS.** Refer to Fig. 3 for exploded view. Steering arm (9) is keyed to spindle and retained by a clamp screw. Spindle

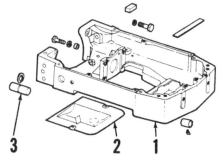

Fig. 2–View of front support casting and associated parts.

1. Front support
2. Pan
3. Rear pivot pin

end play should be maintained at not more than 0.030 by repositioning steering arm on shaft. Tighten steering arm clamp screw to approximately 85 ft.-lbs. Bushings and associated parts are pre-sized.

TIE RODS AND TOE-IN

All Models

3. The recommended toe-in is ⅛-⅜ inch. Remove cap screw in outer clamp and loosen clamp screw in tie rod end (12—Fig. 3); then turn tie rod tube (11) as required. Both tie rods should be adjusted equally.

On manual steering models, stops on both spindles (1) should contact stops on axle extensions (7) at the same time. On power steering models, none of the spindle stops must contact at extreme turning position. Readjust tie rods if necessary until the correct condition is attained. Tighten tie rod clamp screws to a torque of 70 ft.-lbs. when adjustment is correct.

MANUAL STEERING GEAR

Steering gear is a recirculating ball nut type and the housing is mounted on top side of clutch housing. See Fig. 4 for a cross-sectional view of the manual steering gear unit.

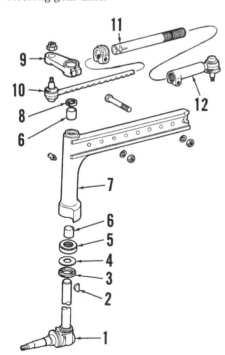

Fig. 3–Axle extension and associated parts of the type used on adjustable axle models. Spindle is similar on fixed tread units.

1. Spindle	7. Axle extension
2. Woodruff key	8. Upper seal
3. Lower seal	9. Steering arm
4. Washer	10. Tie-rod end
5. Thrust bearing	11. Tube
6. Bushing	12. Tie-rod end

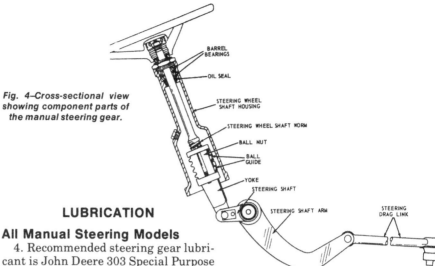

Fig. 4–Cross-sectional view showing component parts of the manual steering gear.

LUBRICATION

All Manual Steering Models

4. Recommended steering gear lubricant is John Deere 303 Special Purpose Oil or Automatic Transmission Fluid, Type A. Fluid should be maintained at level of fill plug located on rear of steering column housing as shown in Fig. 5. Drain plug is located on right hand side of transmission housing

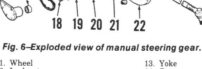

Fig. 5–Steering gear lubricant should be maintained at level of filler plug as shown.

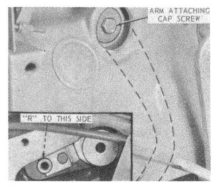

immediately below steering shaft cover.

ADJUSTMENT

All Manual Steering Models

5. Except for tie rods and toe-in (paragraph 3) no adjustment is provided. Excessive steering column shaft end and side play can be eliminated by loosening locknut (2—Fig. 6) and turning adjustor (4) until end play is eliminated and barrel bearings (7) slightly preloaded.

OVERHAUL

All Manual Steering Models

6. **REMOVE AND REINSTALL.** Either the steering wheel shaft unit or rockshaft assembly can be removed independently of the other if required for service. First drain steering gear as outlined in paragraph 4; then refer to the appropriate following paragraphs.

7. **STEERING WHEEL SHAFT HOUSING.** If tractor is equipped with foot throttle, disconnect pedal linkage and lift foot pedal up out of way. On all models, unbolt and remove steering shaft cover (14—Fig. 6). Refer to Inset, Fig. 7 and turn steering wheel until yoke pin is centered in opening as

Fig. 6–Exploded view of manual steering gear.

1. Wheel	13. Yoke
2. Locknut	14. Cover
3. Oil seal	15. Bushing
4. Bearing adjuster	16. Gasket
5. Housing	17. Pin
6. Gasket	18. Cap screw
7. Bearing	19. Clip
8. Thrust washer	20. Cross shaft
9. Retainer	21. Seal
10. Oil seal	22. Steering arm
11. Retainer	23. Cap screw
12. Shaft & ball nut	24. Drag link

Fig. 7–Removing steering cross shaft. Inset shows yoke arm pin, which should be installed with "R" mark visible.

shown; then remove cap screw (18—Fig. 6), clip (18) and yoke pin (17). NOTE: Yoke pin has tapped hole which will accept one of the cover cap screws for pulling leverage.

Remove steering wheel using a suitable puller. Remove the cap screws securing steering shaft housing to clutch housing and dash, and lift unit from tractor.

Install by reversing the disassembly procedure. Make sure "R" mark on yoke (13) is toward cover opening when pin is installed. Tighten lockplate screw (18) to 100 inch-pounds, steering wheel nut to 50 ft.-lbs. and housing cap screws to 35 ft.-lbs. Refill steering gear housing as outlined in paragraph 4.

8. STEERING CROSS SHAFT. The steering cross shaft (20—Fig. 6) can be removed without removing steering wheel shaft, or with wheel shaft and housing removed.

If steering wheel shaft housing is not removed, drain the unit, remove cover (14), cap screw (18) and yoke pin (17) as outlined in paragraph 7; and turn steering wheel until yoke (13) is withdrawn from shaft arm.

Refer to Fig. 7 and remove button plug from left side of clutch housing and remove arm attaching cap screw (23—Fig. 6). Make sure cross shaft arm is still aligned with right cover opening as shown in Inset—Fig. 7 and bump steering shaft to right out of clutch housing and arm.

Install by reversing removal procedure. Tighten arm attaching cap screw (23—Fig. 6) fully, strike arm with a hammer, then retighten to 170 ft.-lbs. Complete the assembly by reversing disassembly procedure and fill steering gear as outlined in paragraph 4.

9. OVERHAUL. To disassemble the removed steering shaft housing, loosen locknut (2—Fig. 6) and remove bearing adjuster (4). Pull steering shaft upward until bearings (7) are exposed and remove upper bearing, retainer (9) and thrust washer halves (8). Lift off lower bearing (7) then withdraw steering shaft downward out of housing and oil seal (10). Yoke (13) can be removed from shaft ball nut by removing the two retaining cap screws. Do not disassemble ball nut as unit is available only as an assembly.

Install oil seal (10) if necessary, with lip toward inside and outer face flush with bottom of seal bore chamfer.

Reassemble by reversing the disassembly procedure. Tighten steering shaft bearing adjuster (4) until a pull of approximately 15 lbs. is required at end of yoke (13) to rock the shaft in barrel bearings (7).

Bushings (15) in cover (14) and clutch housing should be installed 0.030 below face of bushing bore if renewed. Normal clearance between shaft (20) and bushings is 0.005-0.009.

POWER STEERING SYSTEM

The power steering system of series 1020 and 2020, prior to tractor serial number 45200, is an open center valve type and consists of a separate gear type pump, a valve and steering cylinder and steering linkage.

The power steering system of series 1020 and 2020, tractor serial number 45200 and up, is a closed center type having the same basic mechanical components as the earlier open center type. However, the separate gear type pump is not used and the pressurized oil is furnished by the main hydraulic pump via a priority type pressure control valve which is bolted to lower right side of clutch housing.

The power steering system of the series 2020 Low Profile tractors is a closed center, hydrostatic type, consisting of a control valve, metering valve and one double acting hydraulic steering cylinder which controls the front wheels. No mechanical linkage is used between steering wheel and the front wheels. Pressurized oil for operating the system is furnished by the main hydraulic pump and in case of a hydraulic failure, or engine stoppage, steering can still be accomplished with the oil that is trapped within the power steering system.

The power steering system of Series 1520, 1530 and 2030 tractors is closed center type similar in all details to that used in late series 1020 and 2020 tractors.

TROUBLE SHOOTING
Except Series 2020 Low Profile
10. Any problems that may develop in the power steering system usually appear as sluggish steering, power steering in one direction only, or excessive noise in the power steering pump.

Sluggish steering can usually be attributed to:
a. Defective steering valve seat "O" rings which usually produces slow steering in one direction only.
b. Power steering pump not maintaining flow or pressure. (early models)
c. Insufficient oil supply
d. Clogged filter
e. Faulty steering piston "O" rings

Power steering in one direction only can usually be attributed to:

a. Defective steering valve seat "O" rings.
b. Piston retaining snap ring dislodged (left turn only).

Excessive power steering pump noise (early models) can usually be attributed to:
a. Faulty pump bearings or other mechanical damage.
b. Air leak on inlet side of pump.
c. Oil foaming.

Series 2020 Low Profile
11. Any problems that develop in the power steering usually appear as hard steering, escessive slippage, slippage in one direction, or erratic action of the front wheels.

Hard steering can usually be attributed to:
a. Faulty column bearings.
b. Faulty or worn rotor and rotor ring assembly.
c. Jammed or stuck valve spool.
d. Damaged valve spool resulting in high leakage.

Excessive slippage can usually be attributed to:
a. Faulty seals and/or back-up rings (usually rotor)
Slippage in one direction only can usually be attributed to:
a. Faulty or dirty pressure plate check ball.
Erratic action of front wheels can usually be attributed to:
a. Sticky or faulty rotor vanes.
b. Broken or jammed rotor vane springs.
c. Worn or damaged piston "O" ring and/or back-up rings.

SYSTEM CHECKS
Except Series 2020 Low Profile
A series of checks can be made to determine the overall condition of the power steering system, or to isolate a malfunctioning component. To make these tests, proceed as follows:
12. CYCLE TIME TEST. With tractor on a smooth hard surface, start engine and run at high idle rpm. Turn steering wheel as rapidly as possible and check the time required from lock to lock. Time from lock to lock in either direction should be two seconds.

If power steering will operate as outlined, the overall condition of the system can be considered satisfactory.

If time from lock to lock is more than two seconds, either the pump or steering valve is faulty and the malfunctioning unit can be located as outlined in paragraph 13 or 14.

13. **PUMP PSI AND GPM. (Open Center System).** To make an accurate pump test requires a hydraulic test unit which incorporates a flow meter in addition to the pressure gage.

To check pump, proceed as follows: Disconnect pump pressure line from steering valve and connect to inlet line of test unit. Secure outlet line of test unit in transmission case oil filler hole. Open control valve of test unit, start engine and operate at about 2500 rpm. Slowly close the test unit control valve until the pressure gage reads 1000 psi, then observe the flow gage which should read 3½-6 gpm. If this condition is met the pump volume can be considered satisfactory. Continue to slowly close test unit control valve until pump relief valve opens. This should occur somewhere between 1100-1700 psi for early pumps with non-adjustable relief valve; or 1375-1525 psi for late pumps with adjustable relief valve and can be determined by the pump flow lessening, and in most cases, by an audible noise.

If pump does not operate as outlined, refer to paragraph 19 for pump service information; however, if pump is satisfactory and the power steering system does not operate correctly, it can be assumed that the trouble is within the steering valve assembly. Refer to paragraph 14.

NOTE: Do not operate power steering pump with relief valve open for any longer than necessary as damage to relief and metering valve assembly could result.

14. **STEERING VALVE PRESSURE.** A pressure check can be made at the steering valve to determine if valves are leaking, and is done as follows: Remove the sheet metal cover from steering valve. Remove the pressure port plug and install a gage of at least 2000 psi capacity. Block tractor front wheels so they cannot turn, then start engine and run at about 2500 rpm. Note gage reading with steering valve in neutral (at rest). Gage should have a nominal reading of 100 psi. Now turn steering wheel in each direction and note gage reading which should show pump pressure of 1100-1700 psi (early), or 1375-1525 psi (late).

A low pressure reading in both directions would indicate a faulty power steering pump. Refer to paragraphs 13 and 17.

A low reading in one direction only would indicate a faulty steering valve assembly. Refer to paragraph 19 or 21 for steering valve service information.

Series 2020 Low Profile

15. The hydrostatic power steering on the series 2020 Low Profile tractors can be checked for leakage and seal condition as follows:

NOTE: Never restrict the oil return line. To do so will result in the steering wheel shaft oil seal being blown.

Start engine and run at about 1500 rpm. Turn steering wheel in either direction until steering knuckle contacts stop on axle knee, then continue to apply turning effort to steering wheel. Steering wheel should not turn (creep) more than 90 degrees in 10 seconds. Turn front wheels to opposite stop and repeat test. If steering wheel does not creep more than 90 degrees in 10 seconds in either direction, the power steering system can be considered satisfactory. If steering wheel does creep more than 90 degrees in 10 seconds in either direction, there is excessive leakage in the system and the steering valve unit (hydramotor) should be renewed and/or serviced.

NOTE: The test given above can also indicate a leaking steering cylinder. To determine whether steering valve unit or steering cylinder is leaking, disconnect and plug one of the steering cylinder lines, then pressurize the disconnected line by turning the steering wheel. No change in the amount of steering wheel creep would indicate that the steering valve unit is at fault, whereas, a decrease in the amount of steering wheel creep would indicate a faulty steering cylinder.

PUMP

Series 1020 and 2020 (Prior S/N 45200)

16. **REMOVE AND REINSTALL.** To remove power steering pump, first drain and remove radiator as outlined in paragraph 95. Loosen fan belt adjustment and move fan belt out of the way. Disconnect inlet, pressure and bleed lines from pump, then unbolt and remove pump from timing gear cover. Note that the mounting cap screw on right side (behind fan belt) is a socket head type.

Reinstall by reversing removal procedure.

17. **OVERHAUL.** Remove cap screws and separate cover (1—Fig. 8) from pump body (11). Lift body (11) from base (13). Remove gear from drive shaft (20), remove woodruff key (16) and pull drive shaft from base (13). Remove idler shaft and gear, then remove retaining ring (14) and remove gear and woodruff key (16) from idler gear shaft (15). Remove remaining retaining ring from idler shaft, if necessary. Remove plug (9) and pull metering valve (7) and valve spring (6) from bore in pump cover. Any further disassembly required will be obvious.

Clean and inspect all parts. Refer to the following data to determine parts renewal.

Metering Valve Spring
    Test data . . . . . 8¾-10¾ lbs. @ 1⅛ in.
    Gear width 0.5902-0.5908
    Gear Diameter 2.0825-2.0835
    Shaft diameter 0.8744-0.8750

When renewing oil seal (21), install with lip of seal toward pump gears. If bushings in pump cover (1) or base (13), are worn, renew cover or base as bushings are not available separately. If new expansion plug (17) is installed, it

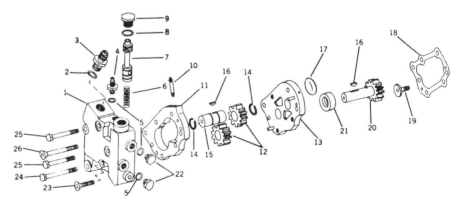

Fig. 8–Exploded view of power steering pump used on models with open center system. Stem (19) is a special thrust screw which screws into cylinder block. Pumps used on tractors, serial number 38520-45199, use "O" rings (not shown) on each side of pump body (11).

1. Pump cover	7. Metering and relief	12. Pump gears
2. "O" ring	valve assembly	13. Pump base
3. Connector	8. "O" ring	14. Retainer ring
4. Connector	9. Plug	15. Idler gear shaft
5. "O" ring	10. Nipple	16. Woodruff key
6. Metering valve	11. Pump body	17. Expansion plug
spring		

18. Gasket
19. Thrust screw
20. Drive gear shaft
21. Oil seal
22. Plug
23. Socket head screw
26. Socket head screw

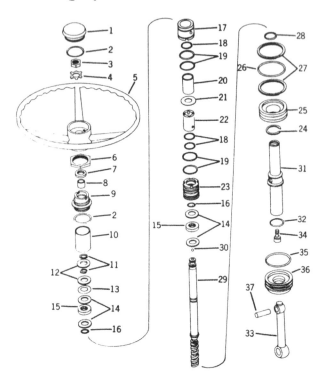

Fig. 9–Series 1020 and 2020 steering valve, prior to serial number 45200, showing component parts and their arrangement.

1. Emblem
2. "O" ring
3. Nut
4. Lock washer
6. Jam nut
7. Oil seal
8. Bushing
9. Adjuster
10. Sleeve
11. Snap ring
12. Washer
13. Shim pack
14. Thrust washer
15. Thrust bearing
16. "O" ring
17. Upper valve inlet seat
18. "O" ring
19. "O" ring
20. Upper valve return seat
21. Valve disc
22. Lower valve return seat
23. Lower valve inlet seat
24. Snap ring
25. Piston
26. "O" ring
27. Back-up ring
28. "O" ring
29. Steering wheel shaft
30. Steel ball
31. Piston rod
32. "O" ring
33. Connecting rod
34. Stop screw
35. "O" ring
36. Piston rod guide
37. Pin

should be flush with surface of bore. Inspect metering and relief valve assembly for damage or obstructions and be sure relief ball is properly seated. Renew complete assembly if necessary as parts are not available separately.

NOTE: As no pump gaskets are used in pumps prior to tractor serial number 38520, it is imperative that the mating surfaces of cover (1), body (11) and base (13) be free of defects as these surfaces provide sealing for the pump. Later pumps use an "O" ring on each side of pump body (11).

In addition, when pump is off, inspect the thrust screw (19) which is located in the cylinder block directly below the camshaft gear (see T—Fig. 30). Renew thrust screw if excessively worn or otherwise damaged.

When reassembling, use Lubriplate on lip of oil seal, lubricate gears and shafts and apply two or three drops of Loctite on threads of pressure line connector (4—Fig. 8), if it was removed and allow four hours for Loctite to cure before exposing connector to any oil.

STEERING VALVE

Series 1020-2020
(Prior S/N 45200)

18. **REMOVE AND REINSTALL.** To remove the steering valve assembly, first remove cap screws from steering shaft cover (3—Fig. 10) and drain oil from steering shaft compartment. If tractor is equipped with a foot accelerator, disconnect rod from foot pedal and swing pedal out of the way. Disconnect return line from steering shaft cover

and pressure line from steering valve housing. Remove steering shaft cover. If necessary, center the steering shaft yoke in cover hole, then remove cap screw (7), lock washer (8) and pin retainer (9). Thread a ⅜-inch cap screw into end of pin (6) and remove pin.

Remove steering wheel emblem, straighten tabs of lock washer and remove steering wheel retaining nut. Attach a puller and remove steering wheel. NOTE: Always use a puller to remove steering wheel. Do not drive on upper end of steering shaft. Remove the metal cover from steering valve housing, then unbolt steering valve assembly from dash and clutch housing and lift unit from tractor.

Reinstall by reversing removal procedure, start engine and cycle system several times to purge any air which might be present.

19. **OVERHAUL.** To disassemble steering valve, use Fig. 9 as a guide and proceed as follows: Place unit in a vise with steering wheel end up. Remove lock nut (6), then using a spanner wrench, remove adjuster (9) and seal (7) assembly. Place steering wheel on

steering shaft and turn steering wheel counter-clockwise until stop is reached, then remove steering wheel and pull steering shaft upward as far as possible which will expose steering valve parts.

Slide sleeve (10) from shaft (29), then remove upper snap ring (11). Slide upper thrust washer (12) off and remove lower snap ring (11). Remaining valve parts can now be removed from steering shaft.

With valve parts removed from steering shaft, push steering shaft, piston and piston rod guide assembly out bottom of housing. Remove piston rod guide (36) and piston (25) from piston rod, then press out connecting rod pin (37) and remove connecting rod (33) from piston rod.

NOTE: At this time, it is advisable to determine the condition of the steering shaft and worm (29) and the steering piston rod (31) as well as the tightness of stop screw (34). If the condition of these parts is satisfactory, it is good policy not to remove the stop screw as it must be reinstalled using Loctite and a minimum of three hours allowed for the Loctite to cure (set). Failure to use Loctite and allow sufficient curing time may result in the stop screw loosening during operation.

If stop screw must be removed, it can be done by using a socket through open end of piston rod (31).

Clean and inspect all parts for wear, scoring, burring or other damage. If a new bushing (8) is installed in adjuster (9), use a piloted driver (JDM 618 or equivalent) and install bushing so top end is flush with chamfer in top side of adjuster. Seal (7) is installed in adjuster with lip facing inward. Valve disc (21) should not be dished, worn or grooved in any way. Be especially sure that all oil grooves and passages of valve seats and housing are open and clean.

Use all new "O" rings and backup rings during reassembly. Lubricate all internal parts as they are installed and reassemble steering valve as follows: If disassembled, screw steering shaft (29) into piston rod (31), place two or three drops of Loctite on threads of stop screw (34), then install stop screw in end of steering shaft and tighten se-

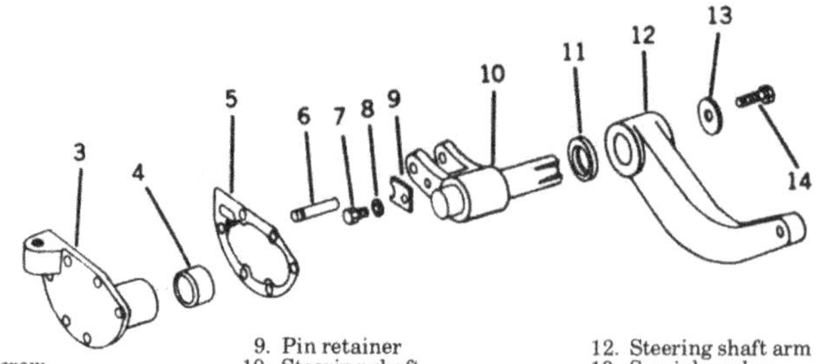

Fig. 10–View of steering shaft component parts which are contained in the top side of clutch housing.

3. Cover
4. Bushing
5. Gasket
6. Pin
7. Cap screw
8. Lock washer
9. Pin retainer
10. Steering shaft
11. Oil seal
12. Steering shaft arm
13. Special washer
14. Cap screw

curely. Allow a minimum of three hours before letting any oil contact stop screw. Be sure all threads are clean and dry before applying Loctite. Place small end of connecting rod (33) into open end of piston rod and install pin (37) through piston rod and connecting rod. Slide piston (25) over piston rod and pin (37) so that recessed portion is facing upward and install snap ring (24).

Before proceeding with further assembly, the shim pack thickness required to determine clearance between steering wheel shaft thrust washer and special washer must be established. To set the valve end play, proceed as follows: Place lower thrust washer and thrust bearing assembly over shaft. Without "O" rings, place lower inlet and return seats over steering shaft with notched ends up. Place valve disc over shaft, then without "O" rings, slide upper inlet and return valve seats over shaft with notched ends down. Install the top thrust washer and bearing assembly and the previously removed shims. Place the two special washers over shims. Now check to see that upper snap ring will fit groove without forcing. Vary the shims under special washers to obtain this condition. See Fig. 11. This will provide 0.000-0.005 clearance between steering shaft thrust washer and special washer when unit is assembled. Keep shim pack intact and pull all parts from steering shaft.

Install "O" ring (26—Fig. 9) with a back-up ring (27) on each side in groove of piston and insert steering shaft and piston rod assembly into housing. Turn steering wheel shaft out of housing as far as possible and install "O" ring (28) in lower end of steering wheel shaft bore in housing. Install first lower thrust washer (14) with chamfer down, thrust bearing (15) and second lower thrust washer (14) with chamfer up. Install the small "O" ring (16) over steering shaft and position against the

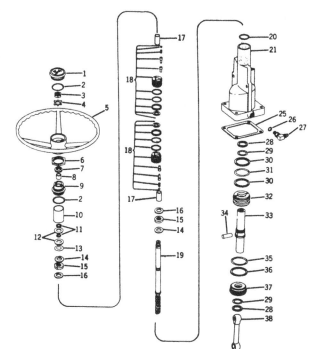

Fig. 12–Exploded view of steering valve used on early models with closed center system, except series 2020 Low Profile. Late models are similar.

1. Emblem
2. "O" ring
3. Nut
4. Lock washer
6. Jam nut
7. Oil seal
8. Bushing
9. Adjuster
10. Sleeve
11. Snap rings
12. Washers
13. Shim
14. Thrust washer
15. Thrust bearing
16. Washer
17. Sleeve
18. Valve body
19. Steering shaft
20. "O" ring
21. Housing
25. Gasket
26. "O" ring
27. Adjustable elbow
28. Back-up ring
29. "O" ring
30. Back-up ring
31. "O" ring
32. Piston
33. Piston rod
34. Pin
35. "O" ring
36. Back-up ring
37. Guide
38. Connecting rod

top of the lower thrust washer. Place lower return valve seat (22) over steering shaft with notched end toward top. Install the two "O" rings (18) in I.D. grooves, and the two "O" rings (19) on O.D. grooves, of lower inlet valve seat (23) and install over lower return valve seat (22) with notched end toward top. Install valve disc (21). Install upper valve return seat (20) with notched end toward valve disc (21), then install small "O" ring (16). Install the remaining "O" ring (18) in I.D. and the two remaining "O" rings (19) on O.D. of upper valve inlet seat and install over upper return valve seat (20) with notched end toward valve disc. Place the lower one of the upper thrust washers (14) over shaft with chamfered side down, then install thrust bearing (15). Install remaining top thrust washer with chamfered side up. Install the previously determined shim pack (13), then install the bottom washer

(special) with cut-out toward shim pack and install lower snap ring (11). Install the remaining top washer (special) with cut-out over bottom snap ring and install the remaining (top) snap ring (11). Turn steering shaft downward into housing and install sleeve (10). Install "O" ring (2) on threaded end of adjuster, coat lip of oil seal (7) with Lubriplate, then either tape steering shaft splines, or use a seal protector, and install adjuster. Use a spanner wrench and tighten adjuster until bottom inlet valve seat butts against shoulder in housing, then back adjuster off ⅛-turn. Hold adjuster in this position then install and tighten lock nut (6). Install small "O" ring (32) in I.D. and large "O" ring (35) on O.D. of piston rod guide (36) and slide guide over piston rod with clutch housing pilot toward outside.

Models With Closed Center System.

20. **REMOVE AND REINSTALL.** Disconnect inlet pressure line, then remove drain plug on right side of clutch housing and drain oil from steering shaft compartment. If tractor is equipped with a foot accelerator, disconnect rod from foot pedal, swing pedal out of the way, then remove the steering shaft cover. If necessary, center the steering shaft yoke in cover hole, then remove cap screw (7—Fig. 10), lockwasher (8) and pin retainer (9). Thread a ⅜-inch cap screw into end of pin (6) and remove pin.

Turn steering wheel from stop to stop to clear oil from steering valve. Remove steering wheel emblem,

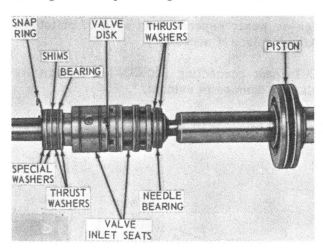

Fig. 11–Refer to text for method of determining the number of shims required during assembly of steering valve. See Fig. 13 for late type valve.

straighten tabs of lockwasher and remove steering wheel retaining nut. Attach a puller and remove steering wheel. Note: Always use a puller to remove steering wheel. DO NOT drive on upper end of steering shaft. Remove metal cover from steering valve housing, then unbolt steering valve assembly from dash and clutch housing and lift unit from tractor.

Reinstall by reversing removal procedure, start engine and cycle system several times to purge any air which may be present.

21. **OVERHAUL.** To disassemble steering valve, use Fig. 12 as a guide and proceed as follows: Place unit in a vise with steering wheel end up. Remove lock nut (6), then using a spanner wrench, remove adjuster (9) and oil seal (7). Pull sleeve (10) from housing. Place steering wheel on shaft and turn shaft clockwise until piston contacts upper end of cylinder, then restrict connecting rod from turning or moving downward and turn steering shaft counterclockwise until steering shaft disengages from the piston steering rod. Remove steering shaft and valve assembly and the piston and piston rod guide from housing.

NOTE: Upper and lower steering valve assemblies are factory assembled and adjusted. If units are disassembled, use caution not to intermix upper and lower valve components. To do so could result in a system malfunction.

To disassemble control valve on all 1020, 1520 and 2020 models; and 2030 models before Serial No. 187301, remove upper snap ring (11) and upper thrust washer (12). Remove lower snap ring (11), then remove remaining valve parts from steering wheel shaft (19).

On all 1530 and late 2030, the retaining snap ring is spring loaded. Remove as shown in Fig. 13, remove special tools then remove valve assembly. Keep parts in proper order as they are removed.

Use a spanner wrench, or strap wrench, and remove piston from piston rod. Connecting rod pin (34—Fig. 12) can now be pressed out and the connecting rod removed from piston rod.

Clean and inspect all parts for wear, scoring, burring or other damage. If a new bushing (8) is installed in adjuster (9), use a piloted driver (JDM618 or equivalent) and install bushing so top end is flush with chamfer in top side of adjuster. Seal (7) is installed in adjuster with lip facing inward.

Use all new "O" rings and back-up rings during reassembly. Lubricate all internal parts as they are installed and reassemble steering valve as follows: Insert connecting rod, small end first, into lower end of piston rod and install pin (34). Pin is pressed in until ends are flush with outer surface of piston rod. Screw piston on piston rod with dowel holes on top side, tighten to 250 ft.-lbs. torque, then install piston "O" ring (31) with back-up ring (30) on each side. Install "O" ring (29) and back-up ring (28) in I.D. of piston rod guide, with back-up ring on bottom side, and install piston rod guide on piston rod with clutch housing pilot toward outside. Install "O" ring and back-up ring in piston rod bore in housing (21) with back-up ring on top side. Install piston, piston rod and piston rod guide assembly in housing and push piston into housing as far as possible.

On early models, valve assembly thrust bearings are pre-loaded 0.001-0.003 by selective fitting of thrust washer shims as shown at (13—Fig. 12). Shims are available in thicknesses of 0.002 and 0.005. To determine shim pack thickness, proceed as follows:

Place lower thrust bearing assembly on steering shaft and be sure chamfer on top thrust washer (bearing race) is toward lower valve body. Install lower operating sleeve, lower valve body and lower spacer and shims. Install the two special washers over shaft, then install upper operating sleeve, upper spacer and shims and upper valve body. Install the upper thrust bearing as-

sembly on steering shaft and be sure chamfer on lower thrust washer (bearing race) is toward upper valve body. Place shim pack over steering shaft (usually 8-0.005 and 2-0.002 shims used) and install the first special upper washer. Hold special washer against shims and vary the number of shims (if necessary) until first snap ring will just fit in its groove, then deduct enough shims to provide 0.001-0.003 preload. Install first snap ring, then install second special washer so cutout in I. D. fits over first snap ring and install second snap ring.

On late models, valve assembly thrust bearings are spring loaded as shown at (B). Install snap ring as shown in Fig. 13.

On all models, install steering shaft and valve assembly into housing, screw steering shaft into piston rod and install upper spacer (10—Fig. 12). Lubricate bushing (8) and coat lips of seal with Lubriplate. Use tape over shaft splines, or a seal protector, then install adjuster and tighten to 35 ft.-lbs. torque. Install and tighten lock nut (6).

Series 2020 Low Profile

22. **REMOVE AND REINSTALL.** To remove the steering valve unit, first remove the operators shield and the cowl cover. Remove the steering wheel emblem, bend down tabs of lock washer and remove steering wheel retaining nut. Attach a puller and remove steering wheel.

NOTE: Always use a puller to remove steering wheel. DO NOT drive on steering shaft.

Clean oil line connections, then disconnect and plug oil lines. Unbolt and remove steering valve unit from mounting bracket. Remove steering shaft tube from steering valve by spreading it slightly at the notch in the tube.

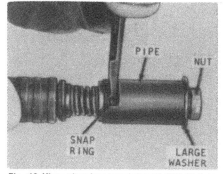

Fig. 13–View showing special tool JDH-41-1 for removing snap ring. Refer to text.

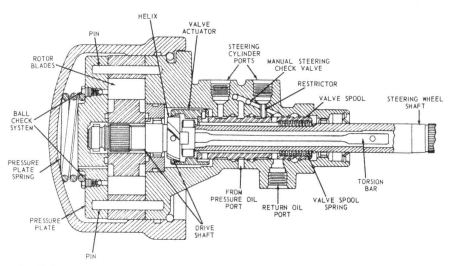

Fig. 15–Cross sectional view of hydrostatic steering unit used only on 2020 Low Profile tractors.

Reinstall by reversing the removal procedure and cycle system several times to purge any air which might be present.

23. **OVERHAUL.** To disassemble the removed Hydramotor, refer to Figs. 15 and 16 and remove cover retaining ring (17—Fig. 16). To remove the retaining ring, drive a 1/8-inch punch into the hole provided in cover (16) to unseat end of ring from groove. With punch under the ring, use screwdriver to pry ring from cover. See Fig. 17.

Place the housing assembly in a vise so that steering shaft is pointing downward. Usually, spring (13—Fig. 16) will push cover from the housing assembly. If binding condition exists, it may be necessary to bump cover loose by tapping around edge of cover with a soft faced mallet.

Remove pressure plate spring (13), then lift off pressure plate (9). Remove dowel pins (24), then using suitable snap ring pliers and screwdriver, remove snap ring (7) from torsion shaft (29). Discard snap ring (7) as a new snap ring must be used when reassembling. Pull pump ring and rotor assembly (6) off torsion shaft (29). Tap end of shaft (28) with a soft faced mallet until bearing support (2) can be removed, then carefully withdraw actuator assembly from housing (22). See Fig. 18. NOTE: It is recommended that the actuator assembly not be disassembled as it is a factory balanced unit.

Housing (22—Fig. 16) and actuator assembly, which includes spring (23), spool (25), actuator (26), shaft (28) and torsion shaft (29) are not serviced separately. If these parts are serviceable, needle bearing (21) and seals (19 and 20) can be renewed as necessary. Install new needle bearing (21) by pressing on lettered side of bearing cage only until bearing cage is flush with counterbore.

Needle bearings (1 and 8) in bearing support (2) and pressure plate (9) may be renewed if support and/or plate are otherwise serviceable. Install new bearings by pressing only on lettered side of bearing cage. Remove plugs (12) and withdraw springs (11) and check balls (10). Inspect the ball seats and balls for excessive wear. Renew parts as necessary.

Rotor, ring, vanes and vane springs are serviced only as a complete assembly (6); however, the unit may be disassembled for cleaning and inspection. Reassemble by placing rotor in ring on flat surface. Insert vanes (rounded side out) in rotor slots aligned with large diameter of ring, turn rotor 1/4-turn and insert remaining vanes. Hook the vane springs behind each vane as shown in Fig. 19, be sure that vane springs are in proper place on both sides of rotor.

To reassemble Hydramotor unit, place housing, with needle bearing, seals and snap ring installed, in a vise with flat (bottom) side up. Check to be sure that pin in actuator is engaged in valve spool; if spool can be pulled away from acuator as shown in Fig. 20, push

spool back into actuator and engage pin into hole in spool. Carefully insert actuator assembly into bore of housing. Install bearing support (2—Fig. 16), with bearing (1), "O" rings and Teflon rotor seal, over end of shaft and care-

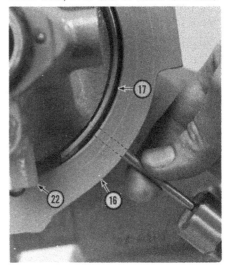

Fig. 17–Cover retaining snap ring can be unseated using a small punch.

Fig. 18–Bearing support and stub shaft assembly can be removed as shown. Repair parts for stub shaft assembly are not available.

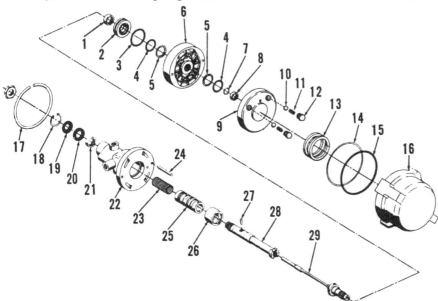

Fig. 16–Exploded view showing component parts of the hydrostatic steering unit.

1. Needle bearing	8. Needle bearing	15. "O" ring	22. Housing
2. Bearing support	9. Pressure plate	16. Cover	23. Valve spool spring
3. "O" ring	10. Ball	17. Retaining ring	24. Dowel pin
4. "O" ring	11. Spring	18. Snap ring	25. Valve spool
5. Rotor seal	12. Plug	19. Dust seal	26. Actuator
6. Rotor and ring	13. Pressure plate spring	20. Oil seal	27. Pin
assembly	14. Back-up ring	21. Needle bearing	28. Shaft
7. Snap ring			29. Torsion shaft

Fig. 19–Assembled view of rotor ring, rotor and vanes. Note installation of vane springs.

fully push the assembly in flush with housing. Place the pump ring rotor assembly on shaft and housing with chamfered outer edge of pump ring away from housing. Install a new rotor retaining snap ring (7) and insert the dowel pins through pump ring and into housing. Stick the "O" ring and Teflon rotor seal into pressure plate with heavy grease, then install pressure plate on shaft, pump ring and rotor assembly and the dowel pins. Place pressure plate spring (13) on pressure plate. Install new "O" ring (15) and back-up ring (14) in groove in cover, then install cover over the assembled steering unit. To install the cover retaining ring, it is recommended that the unit be placed in an arbor press and the housing be pushed into the cover by a sleeve.

CAUTION: **Do Not** push against end of shaft (28). Place retaining ring over housing before placing unit in press. Carefully apply pressure on housing with sleeve until flange on housing is below retaining ring groove in cover. Note that lug on housing must enter slot in cover. If housing binds in cover, **do not** apply heavy pressure; remove unit from press and bump cover loose with mallet. When housing has been pushed sufficiently into cover, install retaining ring in groove with end gap near hole in cover.

PRESSURE CONTROL VALVE
(Except Low Profile)

Models With Closed Center System.

Tractors with closed center hydraulic system are not equipped with a separate power steering pump. The pressurized oil needed for power steering is furnished by the hydraulic system pump, via a priority type pressure control valve, which is mounted on the lower right side of the transmission case.

24. **TESTING.** If a pressure gage is not available, the pressure control valve operation can be checked as follows: Start engine and run at low idle. Turn steering wheel in both directions and note effort required to turn the steering wheel. Then, while turning the steering wheel in one direction, operate either the rockshaft or a remote cylinder and again note the effort required to turn the steering wheel. There should be no change in the effort required to turn steering wheel. A notable increase in the steering effort indicates a faulty operation of the pressure control valve.

If a faulty valve is indicated, obtain a 3000 psi gage, then test and adjust the pressure control valve as outlined in paragraph 25.

25. Remove the plug, or the pump shut-off valve assembly, if so equipped, located directly opposite of the main hydraulic pump stroke control valve and install a 3000 psi gage. Start engine and run at approximately 1900 rpm, then check the hydraulic operations and the pump stand-by pressure which should be 2200-2300 psi. Readjust the pump stroke control valve to 1500 psi, then attempt to operate either the rockshaft or a remote cylinder. The hydraulic function should not occur, and if it does, the pressure control valve is faulty and should be repaired as outlined in paragraph 26 before proceeding further.

To continue testing of the pressure control valve, do either of the following: Completely lower rockshaft and place the rockshaft control lever in the raise position; or, retract a remote cylinder and place the selective control valve lever in the extend position. With engine running at 1900 rpm, adjust the main hydraulic pump stroke control (raise pressure) until either of the above hydraulic operations previously mentioned occurs at its normal rate. This is the regulating point of the pressure control valve and should be 1700-1800 psi. If pressure is not as stated, disconnect front oil line, remove fitting

(11—Fig. 21) and vary shims (7) as required. Shims are 0.030 thick and one shim will change pressure 35-40 psi.

If pressure control valve cannot be adjusted satisfactorily, remove and service valve as outlined in paragraph 26.

Readjust the main hydraulic pump stroke control valve to obtain the recommended 2200-2300 psi stand-by pressure, then remove gage and reinstall plug or pump shut-off valve.

26. **R&R AND OVERHAUL.** To remove pressure control valve, drain transmission, then disconnect inlet (front) oil line and if equipped with remote hydraulics, the rear (outlet) oil line. Unbolt and remove unit from transmission housing.

With valve removed, remove fitting (11—Fig. 21), then remove spool (9), orifice (8), shims (7) and spring (6) from housing (3). Retain shims for subsequent installation.

Inspect spool and housing for wear, scoring or other damage. New spool diameters are 0.7497-0.7503 at front and 0.7257-0.7263 at rear. Spring has a free length of 4⅝ inches and should test 45-55 pounds when compressed to a length of 3½ inches. Be certain the orifice (8) is not worn or damaged.

When reinstalling, attach oil lines before final tightening mounting bolts.

STEERING CYLINDER
Series 2020 Low Profile

27. **R&R AND OVERHAUL.** To remove the power steering cylinder, first clean cylinder, then disconnect and plug both hoses. Remove nuts from cylinder ends and remove cylinder from tractor.

With cylinder removed, remove female end (22—Fig. 22) from piston rod. Remove end plate (20) and using long nose pliers, remove snap ring (19) from I.D. of cylinder barrel. Complete piston, piston rod and bearing (15) assembly can now be pulled from cylinder barrel. If piston and piston rod as-

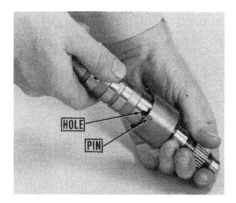

Fig. 20–Pin in actuator sleeve must be engaged in hole in end of control valve spool before actuator assembly is installed. If spool cannot be pulled out of sleeve, pin is engaged.

Fig. 21–Exploded view of pressure control valve used on tractors with closed center system, when equipped with power steering. Line (16) is pressure line for selective control valve.

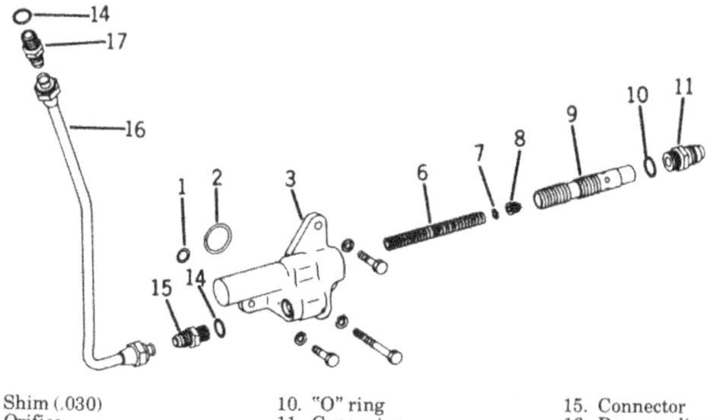

1. "O" ring	7. Shim (.030)	10. "O" ring	15. Connector
2. "O" ring	8. Orifice	11. Connector	16. Pressure line
3. Valve housing	9. Control valve	14. "O" ring	17. Connector
6. Spring			

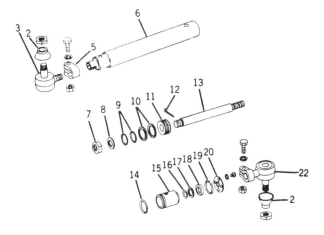

Fig. 22–Exploded view of power steering cylinder used on 2020 Low Profile tractors.

2. Seal
3. Rod end (male)
5. Clamp
6. Barrel
7. Nut
8. Washer
9. "O" rings
10. Piston rings
11. Piston
12. Cotter pin
13. Piston rod
14. "O" ring
15. Bearing
16. "O" ring
17. Back-up washer
18. Seal
19. Snap ring
20. End plate
22. Rod end (female)

sembly is to be disassembled, remove piston from rod and slide bearing assembly off piston end of rod.

Inspect all parts for wear, scoring or other damage. Discard all "O" rings and seals and use new during reassembly.

If either of the Teflon piston rings (10) is damaged, both should be renewed.

Lubricate "O" rings and seals and assemble cylinder as follows: Slide seal (18) on piston end of rod (13) with lip facing toward piston. Install "O" ring (16) and back-up ring (17) in recess at rear of bearing (15) and the large "O" ring (14) on forward end of bearing, then slide bearing on piston end of rod. Slide bearing toward tie-rod end (rear) of piston rod and press oil seal into recess of bearing. Install piston (11) and tighten nut to not more than 50 ft.-lbs. torque and install cotter pin (12). If new piston rings (10) are being installed, installation will be facilitated if piston rings are heated to not more than 170 degrees F. Install piston, piston rod and bearing assembly in cylinder barrel, push bearing in barrel as far as possible, then install snap ring (19), end plate (20) and cylinder ends (3 and 22).

Install steering cylinder on tractor, run engine, then cycle front wheels from lock to lock until all air is purged from steering cylinder.

ENGINE AND COMPONENTS

R&R ENGINE WITH CLUTCH

All Models

28. To remove engine and clutch as a unit, first drain cooling system and if engine is to be disassembled, drain oil pan. Remove front weights if any are installed. Remove the side grille screens and hood. Remove tool box and side rails if so equipped. Remove hydraulic line clamps on right side of tractor and separate the hydraulic pump pressure line at connector located at right rear of front support. Remove the line retainer from lower right front of clutch housing.

Disconnect power steering bleed line from nipple of hydraulic pump inlet line. Loosen fuel line clip at lower right side of front support and disconnect hose from power steering pump return line. Disconnect fuel sender unit wire at connector near fuel pump. Shut off fuel, disconnect fuel outlet line from fuel pump, then unbolt fuel pump from engine. Disconnect bleed lines from top of hydraulic oil reservoir. Separate the hydraulic pump drive coupling and disconnect drag link. Support tractor, attach hoist to front support and axle assembly and separate front support from engine.

Disconnect battery cables and remove batteries. Free bleed line from rocker arm cover. Disconnect wiring from alternator and fuel gage wire from sender connector. Disconnect battery cable and wiring from starter solenoid and wire from oil pressure sending unit. On gasoline models, disconnect primary wire from ignition coil. Disconnect tachometer cable from clutch housing and oil line from power steering valve (if so equipped). Disconnect cold weather starting aid line from intake manifold on diesel models. Remove coolant temperature bulb from water outlet manifold. Disconnect throttle rod and choke cable from carburetor on gasoline models, or disconnect control rod from injection pump on diesel models. If tractor has an underslung exhaust, disconnect exhaust pipe from exhaust manifold. Install two JD-244 engine lifting adapters (or equivalent) to cylinder head and attach hoist to adapters. Remove the battery insulator (wood block), then remove the cap screws which secure cowl (dash) to flywheel housing. Keep engine horizontal with hoist, unbolt engine from clutch housing and pull engine forward until clutch clears clutch shaft.

Reassemble tractor by reversing the disassembly procedure. On tractors with continuous running pto, the pto shaft mates with the pto clutch disc before transmission shaft mates with main clutch disc. If difficulty is encountered while joining engine to clutch housing, bar over engine until both shafts are indexed with both clutch discs and flywheel housing is snug against clutch housing before tightening the retaining cap screws.

Tighten the engine to clutch housing and the front support to engine cap screws to 170 ft.-lbs. torque.

CYLINDER HEAD

Gasoline Models

29. To remove cylinder head, first drain cooling system and remove hood. Disconnect battery ground straps. Disconnect and remove air cleaner tube assembly. Unbolt manifold assembly from cylinder head and if tractor has an underslung exhaust, pull manifold assembly off studs and let exhaust pipe support it. Remove rocker arm vent tube. Disconnect spark plug wires and remove spark plugs. Unbolt coil from cylinder head and let it hang on wires. Disconnect water outlet elbow from cylinder head. Disconnect the coolant temperature bulb. Remove tappet cover and rocker arms and shaft assembly. Identify and remove push rods, then unbolt and remove cylinder head.

When reinstalling cylinder head use a thin coat of Permatex No. 3 on both sides of gasket and tighten the cap screws in sequence shown in Fig. 23 to a torque of 110 ft.-lbs. Be sure oil holes in rear rocker arm bracket and in cylinder head are open and clean as this passage provides lubrication for the rocker arm assembly. Head bolts should be retorqued after engine has run about one hour at 2500 rpm under halfload. Loosen the head bolts about 1/6-turn before retightening them to 110 ft.-lbs. torque. Valve tappet gap (cold) is 0.014 for intake and 0.022 for exhaust.

Diesel Models

30. To remove cylinder head, first drain cooling system and remove hood. Disconnect battery ground straps. Remove air cleaner tube. Remove exhaust manifold from cylinder head and if tractor is equipped with underslung exhaust, the manifold can be left at-

tached to the exhaust pipe. Disconnect injector leak-off line from fuel tank, injectors and injection pump and remove complete leak-off line. Disconnect pressure lines from injectors, remove hold-down clamps and spacers and withdraw injectors. Disconnect the cold starting aid from the inlet manifold elbow and remove elbow. Unbolt the fuel filters from the cylinder head. Plug all fuel openings. Remove vent tube from rocker arm cover, then remove cover and the rocker arm assembly. Identify and remove push rods and the valve stem caps. Disconnect fan baffle and water outlet elbow from cylinder head, then unbolt and remove cylinder head.

When reinstalling cylinder head, use a thin coat of No. 3 Permatex on both sides of head gasket and tighten head bolts to 110 ft.-lbs. torque in sequence shown in Fig. 23. Be sure oil holes in rear rocker arm shaft bracket and cylinder head are open and clean as this passage provides lubrication for the rocker arm assembly. Head bolts should be retorqued after engine has run about one hour at 2500 rpm under half-load. Loosen the head bolts about 1/6-turn before retightening them to 110 ft.-lbs torque. Valve tappet gap (cold) is 0.014 for intake and 0.018 for exhaust.

VALVES AND SEATS

All Models

31. Valves for all models seat directly in the cylinder head. Exhaust valves of the gasoline engines are fitted with Rotocaps while intake valves are fitted with an "O" ring seal. All valves of the diesel engines are fitted on stem end with renewable hardened steel caps.

Specifications for valves and valve seats are as follows:

Gasoline
Face angle .... 44°
Stem diameter (new) .. 0.3715-0.3725
Seat angle .... 45°
Seat width-in. 5/64
Max. seat run-out 0.002
Diesel
Face angle .... 43½°
Stem diameter (new) .. 0.3715-0.3725
Seat angle .... 45°
Seat width-in. 1/16
Max. seat run-out 0.002

Valves are available with 0.003, 0.015 and 0.030 oversize stems. Seats can be narrowed using 20 and 70 degree stones.

TAPPET GAP ADJUSTMENT

All Models

32. Valve tappet gap for all valves can be set with flywheel being placed in only two positions. Valve tappet gap (cold) for all gasoline engines is 0.014 for inlet and 0.022 for exhaust. Valve tappet gap (cold) for all diesel engines is 0.014 for inlet and 0.018 for exhaust.

To set valve tappet gap, turn crankshaft by hand until No. 1 cylinder is at top dead center and TDC timing screw will enter hole in flywheel as shown in Fig. 24. Check the valves to determine whether front cylinder is on compression or exhaust stroke.

NOTE: On three cylinder engines, rear valve (No. 3 exhaust) is open when No. 1 piston is at TDC on compression stroke; on four cylinder engines No. 2 exhaust valve (fourth valve from front) will be partly open.

Refer to the appropriate diagram (Fig. 25, 26, 27 or 28) and adjust the indicated valves; then turn crankshaft one complete turn until timing screw will again enter TDC hole in flywheel and adjust remainder of valves.

VALVE GUIDES

All Models

33. Valve guides are integral with cylinder head and have an inside diameter new of 0.3745-0.3755 which pro-

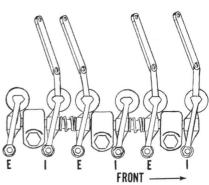

Fig. 24–Reverse timing screw as shown, to find "TDC" timing hole in flywheel, all models. Refer to text.

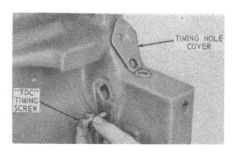

Fig. 25–On three cylinder models the indicated valves can be adjusted when No. 1 piston is at TDC on compression stroke. Turn crankshaft one complete turn and adjust remaining valves; refer to Fig. 26.

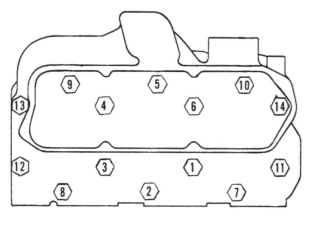

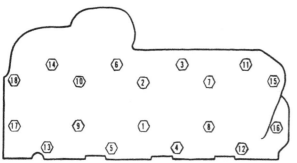

Fig. 23–Tighten cylinder head cap screws to a torque of 110 ft.-lbs., using upper sequence for three cylinder models and lower sequence for four cylinder units.

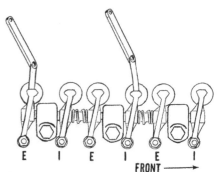

Fig. 26–When No. 1 piston is at TDC on exhaust stroke, the two indicated valves can be gapped. Refer also to Fig. 25.

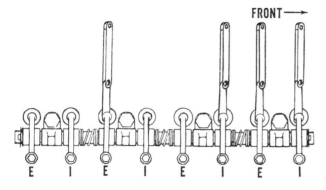

FRONT →

Fig. 27–With No. 1 piston at TDC on compression stroke, gap can be adjusted on the indicated valves. Turn crankshaft one complete turn and adjust remaining valves as shown in Fig. 28.

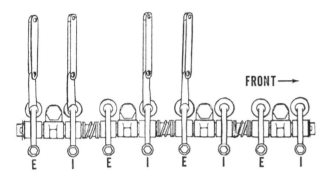

FRONT →

Fig. 28–With No. 4 piston at TDC compression, the four indicated valves may be gapped. Refer also to Fig. 27.

vides 0.002-0.004 operating clearance for valves. Maximum allowable valve stem clearance in guide is 0.006 and when this value is exceeded, ream valve guide as required to fit next oversize valve stem. Valves are available with oversize stem of 0.003, 0.015 and 0.030.

VALVE SPRINGS

All Models

34. Inlet and exhaust valve springs are interchangeable; and in addition, are interchangeable between gasoline and diesel engines. Springs that are distorted, discolored, rusted, or do not meet the following specifications, should be renewed.

Free length (approx.) .... 2⅛ in.
Test lbs. at 1 13/16 in. 52-64
Test lbs. at 1 23/64 in. 129-157

VALVE ROTATORS

All Models

35. Positive type valve rotators are factory installed on the exhaust valves of gasoline engines.

Normal servicing of the rotators consists of renewing the units. It is important however, to check operation of the rotators. If rotator is removed, see that it turns freely in one direction only. If rotator is installed, be sure valve rotates a slight amount each time it opens.

ROCKER ARMS AND SHAFT

All Models

36. Rocker arm and shaft assemblies for all models use identical parts except

for the difference required by the number of cylinders.

Rocker arms are interchangeable and bushings are not available. Inside diameter of shaft bore in rocker arm is 0.790-0.792. Outside diameter of rocker arm shaft is 0.7869-0.7879. Normal operating clearance between rocker arm and shaft is 0.0021-0.0051. Renew rocker arm and/or shaft if clearance is excessive.

Valve stem contacting surface of rocker arm may be refaced but original radius must be maintained.

When reinstalling rocker arm assembly, be sure oil holes and passages are open and clean. Pay particular attention to the rear mounting bracket as lubrication is fed to rocker arm shaft through this passage. Oil hole in rocker arm shaft must face downward when installed on cylinder head.

CAM FOLLOWERS

All Models

37. The cylinder type cam followers (tappets) can be removed from below after camshaft has been removed. If necessary, they can also be removed from above after cylinder head, rocker arm shaft and push rods are removed. Cam followers are available in standard size only and operate directly in machined bores in cylinder block.

It is recommended that new cam followers be installed if a new camshaft is being installed.

VALVE TIMING

All Models

38. Valves are correctly timed when

timing mark on camshaft gear is aligned with timing tool (JD254) when tool is aligned with crankshaft and camshaft centerlines as shown in Fig. 31.

TIMING GEAR COVER

All Models

39. To remove timing gear cover, first remove the front axle and front support assembly as outlined in paragraph 28.

With front support assembly removed, remove fan, fan belt, alternator and water pump. Remove crankshaft pulley retaining cap screw, attach puller and remove pulley. Remove power steering pump, if so equipped. Remove the oil pressure regulating plug, spring and valve (See Fig. 29). Drain and remove oil pan, then unbolt and remove the timing gear cover. See Fig. 30.

Fig. 29–Oil pressure relief valve (except early models) is located as shown. Unit can be adjusted by using shims under forward end of spring. Also see Fig. 43.

Fig. 30–View of four cylinder diesel engine gear train. Three cylinder engines do not use balancer shafts.

B. Balance shaft gear
C. Crankshaft gear
G. Camshaft gear
L. Lower idler gear
O. Oil pump gear
P. Injection pump gear
T. Thrust screw
U. Upper idler gear

With timing gear cover removed, the crankshaft front oil seal can be renewed. To renew oil seal, coat outside diameter of seal with sealing compound and with seal lip toward inside, support timing gear cover around seal area and press seal into bore until it bottoms.

NOTE: Do not attempt to install the front oil seal in timing gear cover without providing support around seal area. Cover is a light cast aluminum alloy and could be warped or cracked rather easily.

On gasoline engines, the governor shaft front bushing is also in the timing gear cover. Inspect this bushing each time timing gear cover is removed and renew bushing, if necessary.

CAMSHAFT

All Models

40. To remove camshaft, timing gear cover must be removed as outlined in paragraph 39. Remove vent tube, rocker arm cover, rocker arm assembly and push rods. Shut off fuel and unbolt fuel pump from cylinder block. On gasoline engines, remove distributor. Use wires with a 90 degree bend in end, pushed into push rod bore of tappet, to hold tappets away from camshaft lobes (wood doweling of proper size and spring type clothes pins can also be used.) Turn engine until thrust plate retaining cap screws can be reached through holes in camshaft gear, then remove cap screws and pull camshaft and thrust plate from cylinder block.

NOTE: If upper idler gear is not being removed, mark camshaft gear and upper idler gear so camshaft can be reinstalled in its original position. If upper idler gear is being removed, turn engine to TDC and align camshaft gear timing mark as shown in Fig. 31 when installing the upper idler gear.

Support camshaft gear, press camshaft from gear and remove woodruff key.

Fig. 31–With engine at TDC and timing tool (TT) positioned on shafts centerline as shown, camshaft gear timing mark (TM) will be directly under edge of timing tool.

If tachometer drive shaft at aft end of camshaft requires renewal thread exposed end and install a nut, then attach a puller to nut and remove the tachometer drive shaft from camshaft.

Camshaft is carried in three unbushed bores in cylinder block. When checking camshaft journal diameters, also check inside diameter of the camshaft journal bores using the following data.

Camshaft Journal O. D. 2.1997-2.2007
Camshaft Journal bore
I. D. 2.2042-2.2052
Normal operating
clearance 0.0035-0.0055
Max. allowable clearance 0.007
Camshaft end play 0.0025-0.0085
Max. allowable end play 0.015
Trust plate thickness (new) 0.156-0.158

When installing new camshaft gear, be sure timing mark is toward front and support camshaft under front bearing journal. When installing new tachometer drive shaft, be sure drive slot is toward rear and support camshaft under rear journal.

When installing camshaft in cylinder block be sure timing mark is aligned as shown in Fig. 31 and tighten thrust plate cap screws to 35 ft.-lbs. torque.

BALANCER SHAFTS

Four Cylinder Models

Four cylinder engines are equipped with two balancer shafts which are located below the crankshaft on opposite sides of the crankcase. Each shaft is carried in three renewable bushings which are located in bores with the cylinder block. The right hand balancer shaft is driven by the lower idler gear and the left hand balancer shaft is driven by the oil pump gear. See Fig. 30. Shafts rotate in opposite directions at twice engine speed and are designed to dampen the vibration which is inherent in four cylinder engines.

41. To remove the balancer shafts, first remove the timing gear cover as outlined in paragraph 39. Remove lower idler gear and oil pump gear.

At this time, check end play of both balancer shafts. End play should be 0.002-0.008 and if end play exceeds 0.015, renew thrust plates during assembly.

Identify balancer shafts as to right and left, unbolt and remove thrust plates and carefully withdraw balancer shafts from cylinder block.

Use the following specifications and check balancer shaft, bushings and thrust plates.

Shaft journal O.D. 1.4995-1.5005
Bushing I. D. 1.5020-1.5040

Shaft operating
clearance 0.0015-0.0045
Max. allowable clearance 0.006
Shaft end play 0.002-0.008
Max. allowable end play 0.015
Thrust plate thickness 0.117-0.119

Renew any parts which do not meet specifications. The two front balancer shaft bushings for either shaft can be renewed with engine in tractor, however, if either of the two rear bushings require renewal, remove the engine, flywheel and flywheel housing. This will permit staking of the rear bushings.

When installing bushings, use a piloted driver (JD-249 or equivalent) and install bushings from front so that front of bushing is flush with chamfer at front of bore and oil holes are aligned with oil holes in cylinder block. With bushings pressed in place, they must be staked as follows: Use John Deere tool number JD-255 and place the half-round portion in I.D. of bushing so the staking ball (B—Fig. 32) is in round relief in bushing groove directly opposite to bushing oil hole. Turn the square half of tool so the correct size dowel (D) is toward lead (dowel) hole and position square half of tool on crankcase boss. Check alignment of cap screw holes and if necessary, turn half-round portion of tool end-for-end. Install cap screws, BE SURE staking ball is in the relief, then tighten cap screws evenly until the half-round half of tool butts against bushing I.D. This will indent the bushing into dowel hole and stake bushing in bore.

If new gears are being installed on balancer shafts, be sure timing mark is toward front and support shaft on both sides of front journal with tool JD-247, or equivalent. Press gear on shaft until it is flush with end of shaft. Be sure gear is within 0.001 of being flush with end of shaft as this controls shaft end play.

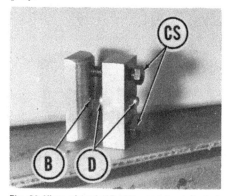

Fig. 32–View of tool (JD-225) used to stake balance shaft bushings. Half-round (left hand) part of tool contains staking ball and is used in I. D. of bushing. Refer to text.

B. Staking ball
D. Dowels

CS. Cap screws

Reinstall balance shafts by reversing removal procedure, however before installing the lower idler gear, set engine on TDC and align crankshaft centerline, timing mark and balance shaft centerline using John Deere timing tool JD-254, or equivalent, as shown in Figs. 33 and 34.

NOTE: When installed and properly timed, keyways of both shafts should be pointing straight up. If not, the shafts have been installed in the wrong bores and must be switched.

IDLER GEARS

All Models

42. All engines are equipped with upper and lower idler gears. Idler gears are bushed and operate on stationary shafts which are attached to the engine front plate with cap screws. Idler gear end play is controlled by thrust washers. Both idler gears are driven by the crankshaft and the upper idler gear drives the camshaft and the diesel injection pump drive gear, or the gasoline engine governor drive gear. The lower idler gear drives the right hand balance shaft gear and the oil pump drive gear. The oil pump drive gear drives the left hand balance shaft gear. See Fig. 30.

To remove idler gears, remove oil pan and timing gear cover, wedge a clean rag between gears, then remove cap screw and pull gear and thrust washers from shaft. Idler gear shaft can now be removed.

Clean and inspect idler gears and shafts and refer to the following specifications.

Shaft O.D. 1.7495-1.7505
Bushing I.D. 1.7515-1.7535
Operating clearance 0.001-0.004
Max. allowable clearance 0.006
End play 0.001-0.007
Max. allowable end play 0.015

Reinstall by reversing removal procedure and be sure camshaft, diesel injection pump and both balance shafts are timed as indicated in Figs. 31, 33, 34 and 35. Tighten shaft cap screws to 85 ft.-lbs. torque.

TIMING GEARS

All Models

43. **CAMSHAFT GEAR.** The camshaft gear (G—Fig. 30) is keyed and pressed on the camshaft. The fit of gear on camshaft is such that removal of the camshaft, as outlined in paragraph 40 is recommended. Camshaft is correctly timed when centerline of camshaft, timing mark on camshaft gear and centerline of crankshaft are aligned and crankshaft is at TDC-1 as shown in Fig. 31.

44. **CRANKSHAFT GEAR.** Renewal of crankshaft gear requires removal of crankshaft as outlined in paragraph 56. Gear is keyed and pressed on crankshaft. Support crankshaft under first throw when installing new gear. Installation of gear may be eased by heating gear in oil.

Crankshaft gear has no timing marks but keyway in crankshaft will be straight up when engine is at TDC.

45. **INJECTION PUMP GEAR AND SHAFT.** Type of gear and method of attachment will depend on what model injection pump is used; refer to Fig. 36.

On Model CDC injection pump, gear can be removed after removing the four retaining cap screws. Gear shaft is an integral part of the pump and screws must be removed in pump removal. Tighten the three outer screws to a torque of 180 inch pounds and the one center screw to 100 inch pounds.

On Model CBC pump, gear and shaft unit can be withdrawn from timing gear housing and injection pump after

timing gear cover is off. Gear is retained by three unevenly spaced cap screws which should be tightened to a torque of 180 inch pounds.

On Models JDB and DBG injection pump, gear and shaft unit can be withdrawn from timing gear housing and pump after cover is off. Gear is keyed to shaft and retained by washer and nut as shown. Tighten nut to a torque of 35 ft.-lbs.

On ROTO DIESEL injection pump, gear can be removed after removing the three retaining cap screws. Gear shaft is an integral part of pump and gear must be removed in pump removal. Dowel pin positions the gear on pump hub. Tighten gear retaining cap screws to a torque of 18 ft.-lbs.

On all models, injection pump drive gear is interchangeable for three and four cylinder models. Two timing marks appear on the gear, each identified by a stamped "3" or "4". Use the timing mark which indicates the number of cylinders, when timing the gears as shown in Fig. 35.

46. **GOVERNOR GEAR.** Refer to paragraph 93 for information on the gasoline engine governor assembly. Governor gear has no timing marks and need not be timed.

47. **TIMING GEAR BACKLASH.** Excessive timing gear backlash may be corrected by renewing the gears concerned, or in some instances by renewing idler gear bushing and/or shaft. Refer to the following for maximum recommended backlash:

Crankshaft gear to upper idler   0.0166
Camshaft gear to upper idler . . 0.0135
Injection pump gear to upper
  idler 0.0135
Crankshaft gear to lower idler . 0.0137
Balance shaft gear to lower
  idler 0.0156
Oil pump gear to lower idler . . . 0.0147
Oil pump gear to balance shaft
  gear 0.0147

Fig. 33–When timing tool (TT) is placed between centerlines of crankshaft and right balance shaft, the timing mark (TM) on gear (B) will be directly below timing tool when shaft is correctly timed.

Fig. 34–When timing tool (TT) is placed between centerlines of crankshaft and left balance shaft, the timing mark (TM) on gear (B) will be directly below timing tool when shaft is correctly timed.

Fig. 35–Injection pump gear (P) is correctly timed if timing mark (TM) is directly below timing tool (TT) when tool is placed between centerlines of crankshaft and injection pump shafts.

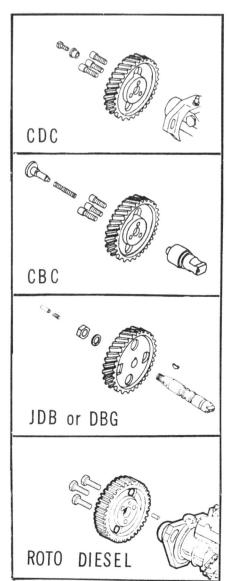

CDC

CBC

JDB or DBG

ROTO DIESEL

Fig. 36–Different injection pumps have been used requiring different gears and removal procedures. Refer to paragraph 45 for details.

ROD AND PISTON UNITS

All Models

48. Piston and connecting rod assemblies are removed from above after removing cylinder head and oil pan. Secure cylinder liners (sleeves) in cylinder to prevent liners from moving as crankshaft is turned.

All pistons have the word "FRONT" stamped on head of piston. Diesel engine connecting rods also have the word "FRONT" stamped (embossed) in the web of connecting rod. Gasoline engine connecting rods and caps have small raised (pip) marks and these "pips" should be in register and towards camshaft side of engine when installed. Replacement rods are not numbered and should be stamped with correct cylinder number. When installing rod and piston units, lubricate

and tighten rod screws to 45 ft.-lbs on 1020, 1520 and 2020 gasoline modesl, 70 ft.-lbs. on 1020, 1520 and 2020 diesel models, 65 ft.-lbs. on 1530 and 2030 gasoline and diesel models.

PISTONS, RINGS AND SLEEVES

All Models

49. All pistons are cam-ground, forged aluminum-alloy and are fitted with three rings located above the piston pin. All pistons have the word "FRONT" stamped on the piston head and are available in standard size only.

Top piston ring on the diesel engines is of keystone design and a wear gage (JDE-62) should be used for checking piston groove wear. On all other rings, normal side clearance is 0.0035-0.005 with a wear limit of 0.008. Installation instructions for piston rings is included in ring kits.

The renewable wet type cylinder sleeves are available in standard size only. Sleeve flange at upper edge is sealed by the cylinder head gasket. Sleeves are sealed at lower edge by packing which may be of the type shown in Fig. 38, 39 or 40. Sleeves normally require loosening using a sleeve puller, after which they can be withdrawn by hand. Out-of-round or taper should not exceed 0.005. If sleeve is to be reused, it should be deglazed using a normal cross-hatch pattern. Sleeves are interchangeable in gasoline and diesel versions of Series 1520 (4.02 inch bore). Gasoline and diesel sleeves are not interchangeable in Series 1020 or 2020 because of the longer stroke in diesel engines.

When reinstalling sleeves, first make sure sleeve and block bore are absolutely clean and dry. Carefully remove any rust or scale from seating surfaces and packing grooves, and from water jacket in areas where loose scale might interfere with sleeve or packing installation. If sleeves are being re-used, buff rust and scale from outside of sleeve.

Install sleeve without the seals and measure standout as shown in Fig. 37. Also check to be sure that sleeve will slip fully into bore without force. If sleeve cannot be pushed down by hand, recheck for scale or burrs. If necessary, select another sleeve. After matching sleeves to all the bores, mark the sleeves then refer to the appropriate following paragraph for packing and sleeve installation.

50. **SQUARE SECTION PACKING TYPE.** Refer to Fig. 38. Install dry packing on cylinder above (liner), making sure packing is not twisted, rolled over or turned inside out in installation. Coat packing and lower step of sleeve liberally with engine oil and immediately install sleeve by pushing straight down without twisting during installation. Fully seat sleeve if necessary, using a hardwood block and hammer.

51. **SQUARE PACKING AND O-RINGS.** Refer to Fig. 39. Install red O-rings (2) dry in block grooves, making sure rings are not twisted or damaged. Lightly oil the machined lower end of sleeve and install black packing (1) on sleeve making sure packing does not twist or turn over during installation. Coat sealing surfaces of block bore (including O-rings) with engine oil then install sleeve by hand, pushing straight down. Seat the sleeve if necessary, using a hardwood block and hammer.

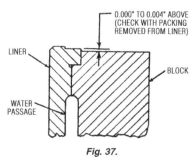

Fig. 37.

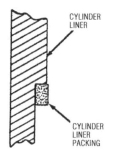

Fig. 38—Cross section of cylinder sleeve, showing square section packing properly installed. Refer to paragraph 50.

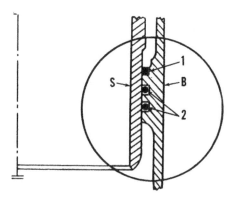

Fig. 39–Correct installation of square packing and O-rings used for sealing cylinder sleeves on some models. Refer to paragraph 51.

1. Square packing	B. Cylinder block
2. O-rings	S. Cylinder sleeve

52. MULTI-SEAL LINER KIT. Refer to Fig. 40. First seal step (A—Inset) should have a 45° lead-in chamfer approximately 0.040 wide. Also check for sharp edges or roughness on lower step (B). Break the corner using emery cloth if edges are sharp or rough.

Upper and lower O-rings (C) are interchangeable. Immerse O-rings in liquid soap solution and install in ascending order (lower ring first). Check to make sure that O-rings are straight and properly positioned. It should be possible to fully install the sleeve (to point of packing compression) by applying the pressure of both thumbs to upper flange.

53. SPECIFICATIONS. Specifications of pistons and sleeves are as follows:

3.86 Bore Engines
Sleeve Bore 3.8581-3.8595
Piston Skirt Diameter:
   Gasoline 3.8561-3.8581
   Diesel 3.8541-3.8551
Piston Skirt Clearance:
   Gasoline 0.0000-0.0034
   Diesel 0.0030-0.0054
4.02 Bore Engines
Sleeve Bore 4.0150-4.0164
Piston Skirt Diameter:
   Gasoline 4.0135-4.0155
   Diesel 4.0130-4.0150
Piston Skirt Clearance: (Selective Fit)
   Gasoline 0.000-0.0024
   Diesel 0.0015-0.0039

NOTE: Manufacturer indicates that zero clearance is apparent, rather than actual, as piston skirt will conform within narrow limits to cylinder size.

PISTON PINS AND BUSHINGS

All Models

54. The full floating piston pins are

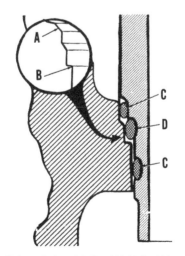

Fig. 40—Installation details of Multi-Seal Liner Kit used in some models; refer to paragraph 52.

A. 45° chamfer   C. Upper & lower rings
B. Sharp edge broken   D. Middle ring

retained in pistons by snap rings. A pin bushing is fitted in upper end of connecting rod and bushing must be reamed after installation to provide a thumb press fit for the piston pin. Piston pin to piston clearance is 0.0001-0.0008.

Piston pin diameter is 1.1886—1.1896 for all 1020, 1520 and 2020 tractors, and for early 1530 and 2030 models. Late tractors use a piston pin 1.376—1.377 inches in diameter.

CONNECTING RODS AND BEARINGS

All Models

55. The steel-backed, aluminum lined bearings can be renewed without removing rod and piston unit by removing oil pan and rod caps.

On 1520 gasoline and all diesel engines, connecting rod big end parting line is diagonally cut and rod cap is offset away from camshaft side as shown in Fig. 41. A tongue and groove cap joint positively locates the cap. Rod marking "FRONT" or raised "pips" should be forward and locating tangs for bearing inserts should be together when cap is installed.

Connecting rod bearings are available in undersizes of 0.002, 0.003, 0.010, 0.012, 0.020, 0.022, 0.030 and 0.032 as well as standard. Refer to the following specifications:
Crankpin Diameter:
   Series 1520 Gasoline 2.7480-2.7490
   Other Gasoline Models2.3085-2.3100
   Diesel Models 2.7480-2.7490
Diametral Clearance:
   Series 1520 Gasoline 0.0012-0.0042
   Other Gasoline Models0.0014-0.0044
   Diesel Models 0.0012-0.0042
Cap Screw Torque—ft.-lbs.
   1020, 1520, 2020
      Gasoline Models 45
   1020, 1520, 2020 Diesel Models . . 70
   1530, 2030 Gasoline and
      Diesel Models 65

CRANKSHAFT AND BEARINGS

All Models

56. Crankshaft in Series 1020, 2020 and 2030 gasoline engines is supported in three main bearings. Series 1520 gasoline and all three cylinder diesel engines have four main bearings while the four cylinder diesel engine has five main bearings. Main bearing inserts may be renewed after removing oil pan, oil pump and main bearing caps. All main bearing caps except rear are interchangeable and should be identified prior to removal so they can be reinstalled in their original position. Rear main bearing is flanged and controls crankshaft end play which should be 0.002-0.008. Install main bearing

caps with previously affixed identification marks aligned and tighten the retaining screws to 85 ft.-lbs. torque.

To remove the crankshaft, it is necessary to remove engine from tractor. With engine removed, remove oil pan, oil pump, cylinder head and the connecting rod and piston units. Remove clutch, flywheel and flywheel housing. Remove timing gear cover, camshaft, injection pump drive gear and shaft and both idler gears. Remove balance shaft thrust plates and the power steering pump thrust screw, then remove engine front plate from cylinder block. Be sure main bearing caps are identified for reinstallation, then remove all bearing caps and lift crankshaft from cylinder block.

Check crankshaft and bearings for wear, scoring or out-of-round condition using the following specifications.
Main journal diameter . 3.1235-3.1245
Crankpin diameter:
   Series 1520 Gasoline 2.7480-2.7490
   Other Gasoline Models2.3085-2.3095
   Diesel Models 2.7480-2.7490
End play, recommended . .0.002-0.008
End play, max. allowable 0.015
Max. allowable journal taper
   (per inch) 0.001
Max. allowable journal out-of-
   round 0.003
Main bearing diametral
   clearance 0.0011-0.0041
   Maximum allowable 0.006

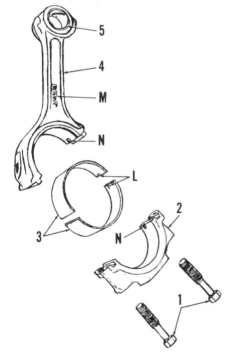

Fig. 41—Connecting rod assembly used in Series 1520 gasoline and all diesel engines.

1. Cap screws   5. Pin bushing
2. Cap   L. Locating tangs
3. Inserts   M. "FRONT" marking
4. Rod   N. Locating notches

If crankshaft does not meet specifications, either renew it or grind to the correct undersize. Main bearings are available in standard size and undersizes of 0.002, 0.003, 0.010, 0.020, 0.022, 0.030 and 0.032.

NOTE: The plug located in aft end of crankshaft is a thrust plug for the transmission input shaft. If plug is damaged in any way, drill it out and press a new plug in until it bottoms.

CRANKSHAFT REAR OIL SEAL

Early Series 1020 and 2020

57. On series 1020 engines before serial number 35895 and 2020 engines before serial number 36138, a bellows type crankshaft rear oil seal is located on crankshaft rear flange and seals against the surface of flywheel housing.

To renew seal, remove engine clutch as outlined in paragraph 113 then remove flywheel and flywheel housing. Seal can now be removed from crankshaft.

NOTE: When installing new oil seal, use of John Deere tool JD-251 (pilot plate and driver) is recommended. See Fig. 42.

To install oil seal, proceed as follows: Place pilot plate, beveled edge rearward over dowel on rear of crankshaft. Position seal so plastic sealing surface is toward rear, then by hand slide seal over pilot plate and start onto crankshaft. Check to be absolutely certain that seal is straight on crankshaft, then slide driver over pilot until it contacts inner metal ring of seal. Bump oil seal on shaft until it bottoms against crankshaft flange. After oil seal is installed, check the bellows action of seal by flexing it gently. Do not bottom the seal too severely or damage to the internal "O" ring could result.

All Other Models

58. On Series 1020 after serial number 35894, 2020 after serial number 36137; and all later models, a lip type crankshaft rear oil seal is contained in flywheel housing and a sealing ring is pressed on mounting flange of crankshaft.

To renew the seal, first remove clutch, flywheel and flywheel housing. If wear ring on crankshaft is damaged, spread the ring using a dull chisel on sealing surface, and withdraw the ring. Install new ring with rounded edge to rear, being careful not to cock the ring. Ring should start by hand and can be seated using a suitable driver such as JD-251. Install seal in flywheel housing working from rear with seal lip to front. When properly installed, seal should be flush with rear of housing bore.

FLYWHEEL

All Models

59. To remove flywheel, first remove clutch as outlined in paragraph 112 through 116, then unbolt and remove flywheel from its doweled position on crankshaft.

To install a new flywheel ring gear, heat to approximately 500 degrees F. and install with chamfered end of teeth toward front of flywheel.

If tractor is a model using a clutch pilot bearing, the bearing can be bumped out of flywheel after snap ring is removed. Install new bearing with shielded side rearward and pack bearing with a high-temperature grease.

When installing flywheel, tighten the retaining cap screws to 85 ft.-lbs. torque.

FLYWHEEL HOUSING

All Models

60. The cast iron flywheel housing is secured to rear face of engine block by eight cap screws. Flywheel housing contains the crankshaft rear oil seal and oil pressure sending unit switch. The rear camshaft bore in block is open and the tachometer drive passes through flywheel housing. It is important therefore, that gasket between block and flywheel housing be in good condition and cap screws properly tightened. Tighten all screws evenly to 22 ft.-lbs., then retorque to 35 ft.-lbs.

OIL PUMP AND RELIEF VALVE

All Models

61. To remove oil pump, first drain and remove oil pan. On four cylinder models, remove timing gear cover as outlined in paragraph 39, then set engine at TDC.

On all models, remove the self-locking nut retaining oil pump drive gear and pull the gear; then unbolt and remove the oil pump.

With pump removed, use Fig. 43 as a guide and proceed as follows: Remove idler gear and drive gear and shaft (11) from pump housing. Check to see that groove-pin (9) is tight in gear and drive shaft. Pin (9) can be renewed if necessary. Check bearing O.D. of drive shaft (11) which should be 0.6295-0.6305. Check I.D. of drive shaft bushing (13) which should be 0.6315-0.6335. Shaft should have a normal operating clearance of 0.001-0.004 in bushing and if clearance is excessive, renew bushing and/or drive shaft and pump gears.

Idler gear shaft (12) can be pressed from pump housing if renewal is necessary. Diameter of new idler shaft is 0.4850-0.4856.

Width of new gears (10) is 1.4163-1.4183 and if gear width does not approximate this dimension, they should be renewed. Install gears and shafts in pump body as shown in Fig. 44 and measure between ends of gear teeth and pump body. This clearance should be 0.001-0.004. Now place a straight

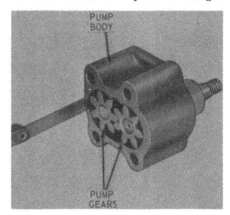

Fig. 44–Clearance between gears and housing should be 0.001-0.004 when measured as shown.

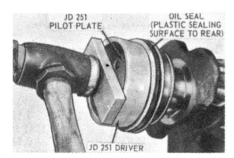

Fig. 42–Installing crankshaft rear oil seal using John Deere tool JD-251. Refer to text.

Fig. 43–Exploded view of engine oil pump used on all engines. Oil outlet tube and relief valve assembly (items 18 through 21) shown are no longer used. See Figs. 29 and 46 for relief valve now being used.

1. Dowel	8. Cover
2. Adapter	9. Groove-pin
3. Idler gear oil tube	10. Pump gears
4. Clamp	11. Drive shaft
	12. Idler shaft
13. Bushing	18. Outlet tube (early)
14. Housing	19. Relief valve (early)
15. Drive gear	20. Spring (early)
16. Nut	21. Cotter pin (early)
17. "O" ring	22. Outlet tube (late)

edge across pump body as shown in Fig. 45 and measure between straight edge and end of pump gears. This clearance should be 0.001-0.006. If either of these clearances are excessive, renew parts as necessary.

NOTE: On early units, a pressure relief valve was incorporated into the oil pump outlet tube as shown in inset of Fig. 43. Later units do not have the relief valve in the pump as the unit has moved to the forward end of the cylinder block oil gallery as indicated by Figs. 29 and 46.

Early pump relief valves were non-adjustable and were designed to maintain 30-40 psi at 2500 engine rpm with engine oil at operating temperature. The later type relief valves can be adjusted by varying shims located between spring and retaining plug. The late type relief valve should be adjusted to maintain 50-60 psi at 2500 engine rpm with engine oil at operating temperature. The valve seat (bushing) for the late type relief valve is pressed into cylinder block as shown in Fig. 46 and is renewable. When in-

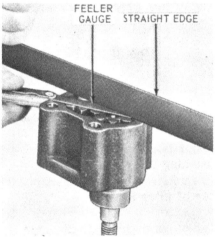

Fig. 45–Clearance between end of gears and cover should be 0.001-0.006 when measured as shown.

Fig. 46–View showing location of the late engine oil pressure relief valve seat. Seat is renewable; refer to text. Also see Fig. 29.

stalling new seat use John Deere tool JD-248, or equivalent tool that will bear only on outer diameter of seat, and press valve seat into cylinder block until outer recessed edge is flush with bottom of counterbore. Do not press on, or otherwise damage, the raised inner rim of valve seat.

Reassemble pump by reversing the disassembly procedure and install pump. On four cylinder models, time the left hand balancer shaft as outlined in paragraph 41. On all models, install the oil pump drive gear. Tighten the gear retaining nut to 35 ft.-lbs. and stake nut to shaft.

CARBURETOR

ADJUSTMENT

All Gasoline Models

62. Initial carburetor settings are 1 turn open for idle mixture and 2 turns open for load adjustment. Both adjustments must be re-set for best performance with engine at operating temperature. Refer to Fig. 47.

OVERHAUL

All Gasoline Models

63. Recommended float setting is ¼-inch, measured from nearest edge of float to gasket surface with throttle body inverted. Both halves of float must be adjusted alike and float must not rub or bind on castings when carburetor is reassembled. If adjustment is required, carefully bend float arms using a bending tool or needle nose pliers.

Carburetor must be disassembled for float adjustment. First clean outside of carburetor with a suitable solvent and remove idle adjustment needle. Remove the four screws which retain throttle body to fuel bowl and lift off

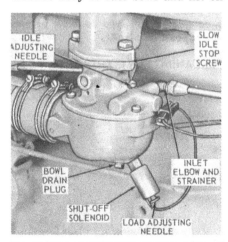

Fig. 47–View of carburetor showing points of adjustment. Note fuel shut-off solenoid.

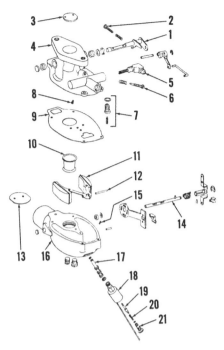

Fig. 48–Exploded view of carburetor of the type used on all gasoline models.

1. Throttle shaft	12. Float shaft
2. Slow idle stop screw	13. Choke valve
3. Throttle valve	14. Choke shaft
4. Throttle body	15. Power jet
5. Inlet strainer	16. Float bowl
6. Idle mixture screw	17. Main nozzle
7. Inlet needle & seat	18. Shut-off solenoid
8. Idle jet	19. Load needle
9. Gasket	20. Spring
10. Venturi	21. Adjusting screw
11. Float	

throttle body, float and venturi as a unit. Remove float shaft and float, then remove and discard gasket, using a new gasket when reassembling. Withdraw and examine inlet needle. If needle is ringed or etched at seating area, renew needle valve and seat assembly before reassembling.

Refer to Fig. 48 for exploded view of carburetor. Check throttle shaft for looseness before or during disassembly and renew the shaft if worn. Remove the plug on blind end of throttle shaft bore and packing from open end before immersing body in cleaning solution; then flush throttle shaft needle bearings thoroughly and relubricate during assembly. The bearings are extremely

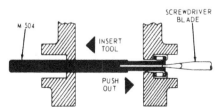

Fig. 49–The special tool M-504 is essential in removing the throttle shaft needle bearings. Hook the lips of split tool behind outer wall of bearing and use a small screwdriver blade as a spreading wedge; then press (do not drive) bearing out in direction shown. The hardened bearing may break if removal is attempted with other tools.

hard and normally last the lifetime of carburetor if properly maintained. Special tools are required to remove and reinstall throttle shaft needle bearings. Refer to Figs. 49 and 50 for tool numbers and procedure. Do not attempt to remove or install the bearings without proper tools.

NOTE: Shut-off solenoid closes load adjustment needle when current to solenoid is interrupted. If malfunction should occur, temporary repairs can be accomplished by interchanging the positions of load needle (19—Fig. 48) and spring (20), causing spring to hold needle OFF it's seat rather than ON seat.

Fig. 50—Use special tool M-503 and a press when installing the hardened throttle shaft needle bearings.

FUEL LIFT PUMP

R&R AND OVERHAUL

All Models

64. Refer to Figs. 51 and 52 for views of lift pumps. The pump should maintain 3.5-4.5 psi pressure 16 inches above pump outlet at 900 engine rpm.

On early models, rocker arm components (14 through 19—Fig. 52) are available in kit form, as are diaphragm (9) and spring (10). Body units (5 and 13) are only available in complete pump assembly. On some late models, only diaphragm, spring and gaskets are available in a repair kit, and in

Fig. 51—Installed view of Airtex fuel pump used on late 2020 and 2030 models.

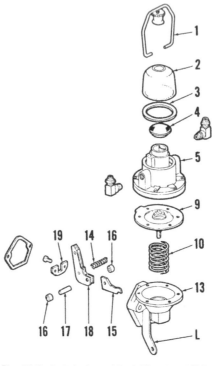

Fig. 52—Exploded view of fuel lift pump of the type used on most early models. Model 1530 pump is similar except that only a repair kit (diaphragm, spring and gaskets) is available for service. Pump shown in Fig. 51 is available only as a complete assembly .

1. Bale	11. Washer
2. Fuel bowl	12. Spring cap
3. Gasket	13. Body
4. Strainer	14. Spring
5. Valve body	15. Diaphragm lever
6. Gasket	16. Bushings
7. Pump valves	17. Pin
8. Retainer	18. Cam lever
9. Diaphragm	19. Clip
10. Spring	L. Primer lever

Airtex fuel pump shown in Fig. 51, pump must be renewed as a complete assembly.

On all models with kits available, mark cover and body before disassembly to assure proper positioning of inlet and outlet ports when reassembling. Install body screws loosely and actuate rocker arm to center and pull the necessary slack into diaphragm. Tighten body screws alternately and evenly.

DIESEL FUEL SYSTEM

FILTERS AND BLEEDING

All Diesel Models

65. **FILTERS.** Tractors may be equipped either with single or dual fuel filters which use a renewable element. Renew the single filter element, or the first stage element of dual filters, every 500 hours of operation. Renew the second stage element of dual filter annually or during major overhaul,

whichever occurs first. However, if tractor has dual stage filters and the first stage was extremely dirty or water soaked, it is recommended that the second stage filter element also be renewed.

Check the glass sediment bowl of single, or first stage of dual filters, for sediment or water, and if necessary, loosen drain plug and operate priming lever of fuel transfer pump to clear deposits from sediment bowl. If tractor has a dual filter and the first stage filter sediment bowl required draining, also drain the aluminum sediment bowl beneath the second stage filter in the same manner.

66. **BLEEDING.** Whenever fuel system has been run dry, or a line has been disconnected, air must be bled from fuel system as follows: Be sure there is sufficient fuel in tank and that tank outlet valve is open. Loosen bleed screw on top of filter (front filter of dual stage) and actuate primer lever of fuel transfer pump until a solid, bubble free stream of fuel emerges from bleed screw, then tighten bleed screw. If tractor has dual filters, repeat operation for second (rear) filter. Loosen pressure line connections at injectors about one turn, open throttle and crank engine until fuel flows from loosened connections, then tighten connections and start engine.

NOTE: If no resistance is felt when operating priming lever of fuel transfer pump and no fuel is pumped, the transfer pump rocker arm is on the high point of pump cam of camshaft. In this case, turn engine to re-position pump can and release pump rocker arm.

INJECTOR NOZZLES

All Diesel Models

WARNING: Fuel leaves the injection nozzles with sufficient force to penetrate the skin. When testing, keep your person clear of the nozzle spray.

67. **TESTING AND LOCATING A FAULTY NOZZLE.** If one engine cylinder is misfiring it is reasonable to suspect a faulty injector. Generally, a faulty injector can be located by running the engine at low idle speed and loosening, one at a time, each high pressure line at injector. As in checking spark plugs in a spark ignition engine, the faulty unit is the one that least affects the engine operation when its line is loosened.

Remove the suspected injector as outlined in paragraph 68. If a suitable nozzle tester is available, test injector as outlined in paragraphs 69 through 73 or install a new or rebuilt unit.

68. REMOVE AND REINSTALL. To remove an injector, remove hood and wash injector, lines and surrounding area with clean diesel fuel. Use hose clamp pliers to expand clamp and pull leak-off boot from injector. Disconnect high pressure line, then cap all openings. Remove cap screw from nozzle clamp and remove clamp and spacer. Pull injector from cylinder head.

NOTE: Unless the carbon stop seal has failed causing injector to stick, the injectors can be easily removed by hand. If injectors cannot be removed by hand, use John Deere nozzle puller JDE-38 and be sure to pull injector straight out of bore. DO NOT attempt to pry injector from cylinder head or damage to injector could result.

Series 1520 diesel engines between Serial Number 95610 and 117499 are equipped with a modified piston T-30015 and injector is raised 0.075 for improved performance. Injector is raised by placing a nylon spacer (1—Fig. 53) on nozzle stem and steel spacer (2) under thick spacer on clamp bolt. It is recommended that the spacers be installed in earlier 1520 tractors when piston (T-30015) is installed. The piston can be identified by the machined peak (Mexican Hat) in center of combustion chamber cavity in piston.

When installing injector, be sure nozzle bore and seal washer seat are clean and free of carbon or other foreign material. Install new seal washer and carbon seal on injector and insert injector into its bore using a slight twisting motion. Install and align locating clamp then install hold-down clamp and spacer and tighten cap screw to 20 ft.-lbs. torque.

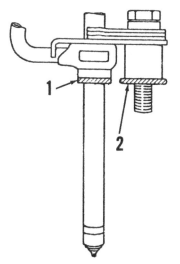

Fig. 53–When modified piston (T-30015) is installed in 1520 diesel engine, injector should be raised 0.075 for best performance. A nylon spacer (1) and steel spacer (2) is provided for this purpose. Refer to paragraph 68.

Bleed system as outlined in paragraph 66 if necessary.

69. TESTING. A complete job of nozzle testing and adjusting requires the use of an approved nozzle tester. Only clean, approved testing oil should be used in the tester tank. The nozzle should be tested for spray pattern, opening pressure, seat leakage and back leakage (leak-off). Injector should produce a distinct audible chatter when being tested and cut off quickly at end of injection with a minimum of seat leakage.

NOTE: When checking spray pattern, turn nozzle about 30 degrees from vertical position. Spray is emitted from nozzle tip at an angle to the centerline of nozzle body and unless injector is angled, the spray may not be completely contained by the beaker. Keep your person clear of the nozzle spray.

70. SPRAY PATTERN. Attach injector to tester and operate tester at approximately 60 strokes per minute and observe the spray pattern. A finely atomized spray should emerge at each nozzle hole and a distinct chatter should be heard as tester is operated. If spray is not symmetrical and is streaky, or if injector does not chatter, overhaul injector as outlined in paragraph 74.

Early 1020 and 2020 tractors used an injector unit (AT18064) having four spray holes and an injection pressure of 2600 psi. Later 1020, 1520 and 2020 tractors and some early 2030 models used a five-hole nozzle. Model 2030 tractors after serial number 211699 and all 1530 units are equipped with a four-hole nozzle which is the recommended service replacement for all five-hole nozzle units.

71. OPENING PRESSURE. The correct opening pressure is 2550-2650 psi for early four-hole nozzles. The first five-hole nozzles (AT 25502) were also set at 2600 psi but later recommendations suggest raising the pressure 200 psi to obtain better performance and longer nozzle life. Later nozzles (AT 32099) used with Model CD pump were factory set at 2800 psi. The five-hole injector (AR 49877) and late four-hole unit (AR 56290) are set at 3000 psi for all applications.

If opening pressure is not correct but nozzle will pass all other tests, adjust opening pressure as follows: Loosen the pressure adjusting screw lock nut (13—Fig. 54), then hold the pressure adjusting screw (14) and back out the valve lift adjusting screw (11) at least one full turn. Actuate tester and adjust nozzle pressure by turning adjusting screw as required. With the correct nozzle opening pressure set, gently

turn the valve lift adjusting screw in until it bottoms, then back it out as follows, to provide the correct valve lift.

On the early four-hole injector nozzle (AT 18064), adjusting screw should be backed out ¾-turn to establish a nominal valve lift of 0.135 inch. On all other nozzles, injector screw should be backed out ½-turn, to establish a nominal valve lift of 0.090 inch.

Hold pressure adjusting screw and tighten lock nut to a torque of 110-115 in.-lbs. for early type adjusting screws, or 70-75 in.-lbs. torque for the later type adjusting screws as shown in Fig. 55.

NOTE: A positive check can be made to see that the lift adjusting screw is bottomed by actuating tester until a pressure of 250 psi above nozzle opening pressure is obtained. Nozzle valve should not open.

72. SEAT LEAKAGE. To check nozzle seat leakage, proceed as follows: Attach injector on tester in a horizontal position. Raise pressure to approximately 2400 psi, hold for 10 seconds and observe nozzle tip. A slight dampness is permissible but should a drop form in the 10 seconds, renew the injector or overhaul as outlined in paragraph 74.

73. BACK LEAKAGE. Attach injector to tester with tip slightly above horizontal. Raise and maintain pressure at approximately 1500 psi and

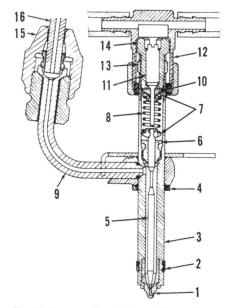

Fig. 54–Cross sectional view of Roosa-Master injector. Nozzle tip (1) and valve guide (6) are parts of finished body and are not serviced separately.

1. Nozzle tip
2. Carbon seal
3. Nozzle body
4. Seal washer
5. Nozzle valve
6. Valve guide
7. Spring seat
8. Pressure spring
9. Inlet fitting
10. Ball washer
11. Lift adjusting screw
12. Boot
13. Lock nut
14. Pressure adjusting screw
15. Compression nut
16. Pressure line

observe leakage from return (top) end of injector. After first drop falls, the back leakage should be 5 to 8 drops every 30 seconds. If back leakage is excessive, renew injector or overhaul as outlined in paragraph 74.

74. **OVERHAUL.** First wash the unit in clean diesel fuel and blow off with clean, dry compressed air. Remove carbon stop seal and sealing washer. Clean carbon from spray tip using a brass wire brush. Also, clean carbon or other deposits from carbon seal groove in injector body. DO NOT use wire brush or other abrasive on the Teflon coating on outside of nozzle body between the seals. Teflon coating can be cleaned with a soft cloth and solvent. Coating may discolor from use, but discoloration is not harmful.

Clamp the nozzle in a soft jawed vise, loosen lock nut (13—Fig. 54) and remove pressure adjusting screw (14), ball washer (10), upper spring seat (7), spring (8) and lower spring seat (7).

If nozzle valve (5) will not slide from body when body is inverted, use special retractor (Fig. 56); or reinstall on nozzle tester with spring and lift adjusting screw removed, and use hydraulic pressure to remove the valve.

Nozzle valve and body are a matched set and should never be intermixed. Keep parts for one injector separate and immerse in clean diesel fuel in a compartmented pan, as unit is disassembled.

Clean all parts thoroughly in clean diesel fuel using a brass wire brush and lint-free wiping towels. Hard carbon or varnish can be loosened with a suitable, non-corrosive solvent.

Clean the spray tip orifices first with an 0.008 cleaning needle held in a pin vise as shown in Fig. 57. On early four-hole nozzles, follow up with a 0.012 needle; on five hole nozzle and late four-hole units, use a 0.010 needle.

Clean the valve seat using a Valve Tip Scraper and light pressure while rotating scraper. Use a Sac Hole Drill to remove carbon from sac hole.

Piston area of valve and guide can be lightly polished by hand, if necessary, using Roosa Master No. 16489 lapping compound. Use the valve retractor to turn valve. Move valve in and out slightly while rotating, but do not apply down pressure while valve tip and seat are in contact.

When assembling the nozzle, back lift adjusting screw (11) several turns out of pressure adjusting screw (14), and reverse disassembly procedure using Fig. 54 as a guide. Adjust opening pressure and valve lift as outlined in paragraph 71, raising pressure 200 psi above that specified if a new spring is installed.

INJECTION PUMP

All Diesel Models

74A. A CAV Roto Diesel injection pump is used on some Model 1530 and 2030 tractors. Roosa Master DBG, CDC, CBC or JDB pumps have been used on other models. All pumps are flange mounted on left side of engine front plate and driven by upper idler gear of timing gear train, but pumps are not individually interchangeable and procedures for removal, installation and adjustment differ slightly. Refer to the following paragraphs for procedure.

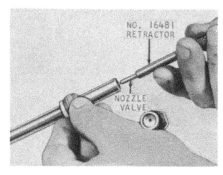

Fig. 56–Use the special retractor as shown, to remove a sticking nozzle valve.

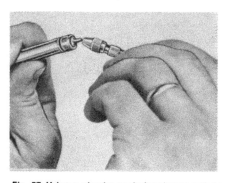

Fig. 57–Using a pin vise and cleaning needle to clean spray tip. Use a 0.008 diameter needle to open holes. Final cleaning should be done with a 0.010 needle on all nozzles except early four-hole type which requires a 0.012 needle.

Injection pump service demands the use of specialized equipment and special training which is beyond the scope of this manual. This section therefore, will cover only the information required for removal, installation and field adjustment of the injection pump.

Model DBG Pump

Series 1020 tractors before serial number 60326 were equipped with injection pump Model DBGFC 33113DH; and 2020 tractors before serial number 50303 with pump model DBGFC 4312DH.

75. **REMOVE AND REINSTALL.** To remove injection pump, first close fuel shut-off valve on fuel tank, then clean injection pump, line connections and the surrounding area with clean diesel fuel. Turn engine in direction of normal rotation until number one piston is starting up on compression stroke. Remove timing pin and cover from flywheel housing, reverse timing pin and reinsert it into threaded hole in flywheel housing. Press on timing pin and continue to turn engine until timing pin enters hole in flywheel. Engine is now at TDC.

NOTE: While it is not absolutely necessary, it is recommended that engine be set at TDC so timing can be checked and adjusted if necessary, when injection pump is reinstalled. If flywheel hole passed timing pin on first attempt, reverse rotation of engine at least ¼-turn before again setting engine TDC.

With engine set at TDC, disconnect fuel inlet line, fuel return line and throttle rod from injection pump. Disconnect pressure lines from injectors and pump and remove pressure lines. Remove pump mounting nuts and pull pump straight rearward off pump shaft.

When reinstalling pump, remove timing hole cover from pump and install timing window. Be sure timing lines on governor weight retainer and pump cam are aligned, start pump on pump shaft, and using seal compressor,

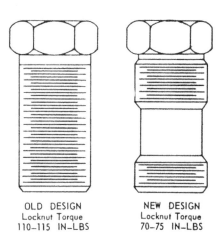

OLD DESIGN
Locknut Torque
110-115 IN-LBS

NEW DESIGN
Locknut Torque
70-75 IN-LBS

Fig. 55–Before tightening lock nut on injector pressure adjusting screw, determine which type is used.

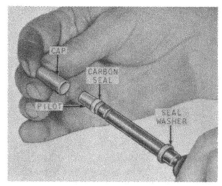

Fig. 58–Use the special pilot or a nozzle storage cap when installing a new carbon seal.

slide pump into position and install the mounting nuts. Rotate pump back and forth, re-align timing marks and tighten mounting nuts.

NOTE: Use caution when sliding injection pump over shaft seals. If undue resistance is encountered, stop and examine rear seal. If lip has been rolled, renew the seal.

Remove timing pin, turn engine two complete revolutions, reinsert timing pin and again check, and align if necessary, the pump timing lines. The normal backlash of timing gears is enough to slightly affect injection pump timing.

Remove timing window, reinstall timing hole cover, throttle linkage and all lines. Bleed fuel system as outlined in paragraph 66.

76. TIMING AND ADJUSTMENT. To check injection timing with injection pump on tractor, first turn engine until number one piston is starting compression stroke, then remove timing pin and cover and reinsert timing pin into threaded hole in flywheel housing. Continue to turn engine until timing pin slides into hole in flywheel.

With engine set at TDC, shut off fuel and remove timing hole cover from injection pump. The timing lines on the governor weight retainer and pump cam should be aligned. If lines are not aligned, loosen pump mounting nuts and rotate pump either way as required to bring the lines into alignment. Hold pump in this position and tighten pump mounting nuts securely.

77. LOAD ADVANCE. The injection pump automatic advance unit is adjusted during production, however, due to variations in both fuel and operating temperatures, it may be necessary to make slight adjustments to provide optimum engine performance. The

pump automatic advance can be checked, and adjusted if necessary, as follows:

NOTE: When working on injection pumps, take into consideration that pumps are sealed and breaking of seals by unauthorized personnel will void the injection pump warranty.

Shut off fuel, remove pump timing hole cover and install timing window (Roosa-Master No. 13366 or John Deere No. JD-259). Identify the window scribe line with scribed line on pump cam ring for future reference. Turn on fuel, start engine and bring to operating temperature. Run engine at 1400 rpm for both 1020 and 2020 series tractors and read the cam advance at timing window. Cam scribe line should be advanced (moved up) 1½ marks on window which equals 3 pump degrees (6 crankshaft degrees) advance. If advance is not correct, use Fig. 59 as a guide and proceed as follows: Remove cap from the advance trimmer screw and loosen lock nut. Use an Allen wrench and turn trimmer screw clockwise to retard timing, or counterclockwise to advance timing.

To complete check, slowly increase engine speed to 2200 rpm for series 1020 tractors, or 2400 rpm for series 2020 tractors, and note the pump cam scribe line which should now have advanced (moved up) 3 marks on window which equals 6 pump degrees (12 crankshaft degrees) advance.

NOTE: At Roosa-Master pump serial number 690393 (1020) or 700034 (2020), the advance piston spring was changed. On these later pumps advance will be 5 degrees at 1400 engine rpm and 8 degrees at 2200 engine rpm (1020) or 2400 engine rpm (2020).

If injection pump will not meet both conditions outlined, either renew or overhaul the pump.

Model CDC Pump

Series 1020 tractors, serial number 60327 to 83747 use Roosa Master Model CDC331-3DG, CDC331-4DG or CDC331-6DG injection pump; Series 1520 tractors before serial number 83759 use Roosa Master Model CDC331-74G injection pump and Series 2020 tractors, serial number 50304 to 84755 use Roosa Master Model CDC431-6DG, CDC431-9DG or CDC431-14DG injection pump. Recommended replacement pump for most models is the appropriate Model CBC pump plus attaching screws, thrust button and spring.

78. REMOVE AND REINSTALL. To remove injection pump, first close fuel shut-off valve on fuel tank, then clean injection pump, line connections and the surrounding area with clean diesel fuel.

Turn engine in the direction of normal rotation until number one piston is starting up on compression stroke, then remove flywheel timing hole cover and timing pin, reverse timing pin and reinsert in the threaded hole. Continue to slowly turn engine until timing pin enters hole in flywheel.

NOTE: If hole in flywheel goes past timing pin, turn engine counterclockwise about ¼-turn, then again turn engine clockwise until timing pin and hole in flywheel are indexed. Engine is now at TDC.

Drain radiator, remove lower radiator hose, then remove access plate from front of timing gear cover. See Fig. 61. Remove pump drive gear mounting screw and the three gear retaining screws. Disconnect throttle rod, fuel supply line and fuel return line from pump. Remove the injector lines, then cap or plug all fuel openings. Support pump, remove the retaining nuts and washers and pull injection pump from engine. Pump

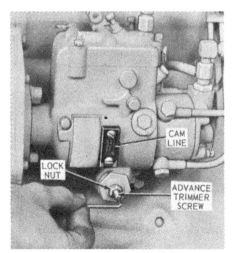

Fig. 59–View showing Model DBG injection pump cam advance trimmer screw. Refer to text for adjustment procedure.

Fig. 60–View showing Roosa-Master Model C injection pump mounted on a series 2020 tractor. Models for the series 1020 tractors are similar except they are for three cylinder engines.

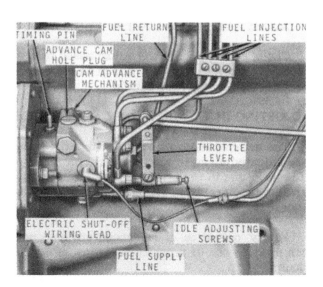

drive gear will be retained by timing gear cover.

When reinstalling pump, be sure engine is positioned at TDC. Remove injection pump timing pin from drive housing, reverse pin and reinsert it in the threaded hole. Turn pump drive shaft until timing pin drops into groove of pump drive shaft.

NOTE: Observe pin carefully as it drops only about 1/16-inch.

Hold pump in this position and mount pump on engine. Install pump drive gear, then recheck pump installation by loosening mounting nuts and turning pump counter-clockwise as far as it will go, then turning pump clockwise until timing pin indexes with pump drive shaft groove. Tighten mounting nuts and complete assembly by reversing disassembly procedure. Bleed fuel system as outlined in paragraph 66.

79. TIMING AND ADJUSTMENT. To check injection timing with injection pump on tractor, first turn engine until number one piston is starting compression stroke, then remove injection pump timing pin from drive housing, reverse timing pin and reinsert it in the threaded hole. Continue to turn

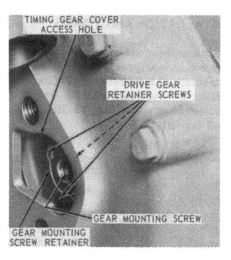

Fig. 61–Remove access hole cover from timing gear cover to gain access to the injection pump drive gear mounting screw and retaining screws on model CDC pump.

engine in normal direction of rotation until timing pin drops into groove of pump drive shaft.

NOTE: Observe timing pin carefully as it will drop only about 1/16-inch when it indexes with groove in pump shaft. If pump shaft groove goes past timing pin, turn engine counter-clockwise about ¼-turn, then again turn engine clockwise.

With injection pump set as outlined, remove flywheel timing pin and timing hole cover, reverse pin and reinsert it in threaded hole. Timing pin should enter hole in engine flywheel. If both timing pins index at the same time, injection timing is correct. See Fig. 62.

If both timing pins do not index at the same time, temporarily remove injection pump timing pin and index the flywheel timing pin. Loosen injection pump mounting nuts, rotate injection pump counter-clockwise on mounting studs and reinsert injection pump timing pin. Slowly turn injection pump clockwise until timing pin indexes with groove in pump drive shaft, then tighten mounting nuts.

80. LOAD (CAM) ADVANCE. The injection pump automatic advance unit is adjusted during production, however, due to variations in both fuel and operating temperatures, it may be necessary to make slight adjustments to provide optimum engine performance. The pump automatic advance can be checked, and adjusted if necessary, as follows: Be sure injection pump is static timed correctly as outlined in paragraph 79, then remove advance cam hole plug and install timing window (part number JD270). See Fig. 63.

NOTE: Notice that timing window has a bulls eye at the center with several concentric circles around it. Each circle is equal to two degrees.

Look through window and locate pump cam pin and align the side of pin nearest to engine with circular line of timing window. This point may vary between individual units and may not be at the bulls eye of timing window. Start engine and bring to operating temperature. Run engine at 800 rpm

Fig. 62–View showing injection pump timing pin and engine timing pin being used to time model CDC injection pump. Refer to text for procedure.

and observe pump cam pin which should still be at zero (original at rest) advance position. Advance engine speed to 1900 rpm and note cam advance in window. Cam pin should have advanced two marks (4 degrees) in window. If advance is not correct, remove cap seal on the trimmer screw, located on engine side of injection pump, and adjust trimmer screw as required. See Fig. JD63.

NOTE: When working on injection pumps, take into consideration that pumps are sealed and breaking of seals by unauthorized personnel will void the injection pump warranty.

Slowly advance engine speed to 2400 rpm and note the cam pin advance which should be 3 marks (6 degrees) in window. If pump does not advance 6 degrees at 2400 rpm, remove pump for test stand adjustment and/or service.

Model CBC Pump

Series 1020 tractors, serial number 83748 to 117499 use Roosa Master Model CBC331-1AL injection pump; Series 1520 tractors, serial number 83760 to 117499 use Roosa Master Model CBC331-3AL injection pump and Series 2020 tractors, serial number 84756 to 117499 use Roosa Master Model CBC431-3AL injection pump. The appropriate model CBC pump is used as parts replacement for the earlier model CDC pump when complete pump is renewed, by also obtaining attaching cap screws, thrust button and spring.

81. REMOVE AND REINSTALL. To remove the injection pump, first close fuel shut-off valve on fuel tank, then clean injection pump, line connections and surrounding area with clean diesel fuel or solvent. Turn engine crankshaft until No. 1 piston is at TDC on compression stroke.

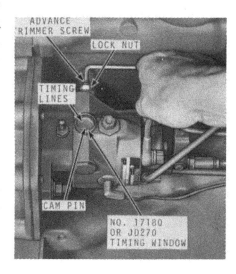

Fig. 63–The cam advance trimmer screw for model CDC pump is located on engine side of injection pump. Note timing window JD270.

NOTE: Pump can be removed and reinstalled without regard to crankshaft timing position, however, positioning crankshaft at TDC-1 is recommended so timing can be properly checked and/or adjusted when pump is reinstalled. If timing is not to be checked, scribe timing marks on injection pump mounting flange and engine front plate which can be realigned when pump is reinstalled.

Disconnect or remove fuel inlet, return and pressure lines, throttle rod and solenoid wire from injection pump. Remove pump mounting stud nuts and pull pump straight to rear off pump shaft.

To install the pump without changing the timing, align mating timing marks (Fig. 64) on drive shaft and pump rotor and reinstall the pump with previously installed scribe lines in register.

To install and time the pump, check to be sure that No. 1 piston is at TDC on compression stroke. Remove timing pin (1—Fig. 65) from top of injection pump body and reinsert the pin long end down. Turn injection pump rotor until timing pin enters timing hole in rotor shaft, then install the pump.

On all installations, tighten injection pump retaining stud nuts to a torque of 30 ft.-lbs. Bleed pump and lines as outlined in paragraph 66. Adjust advance timing if necessary as in paragraph 83 and linkage as in paragraph 89.

82. **TIMING.** To check injection pump timing without removing injection pump, turn engine until No. 1 piston is starting compression stroke, then remove engine timing pin and cover (Fig. 66). Insert timing pin into threaded hole in housing as shown and continue turning crankshaft until timing pin slides into timing hole in flywheel.

Invert pump timing pin (1—Fig. 65). Loosen mounting stud nuts (2) and rotate pump body in slotted holes if necessary until pump timing pin drops into hole in rotor. Tighten pump mounting nuts to 30 ft.-lbs. and reinstall and tighten both timing pins.

83. ADVANCE TIMING. First make sure static timing is properly set as outlined in paragraph 82. Shut off fuel, remove advance cam hole plug and install No. 17180 Timing Window as shown in Fig. 67. The timing window contains a series of concentric circles as shown in inset, and is designed so that all lines are two pump degrees apart. Note the location of hole in center of cam pin, which should be offset from center of bullseye as shown. Turn on fuel and start and warm engine, then note cam advance which should be 5 degrees at 1600 engine rpm. If it is not, loosen the locknut and turn advance trimmer screw until intermediate advance is correct. Maximum advance at full throttle should be 7°. If maximum advance is incorrect or intermediate advance cannot be properly adjusted, renew or overhaul the pump.

Model JDB Pump

Series 1020 tractors after serial number 117500 are equipped with Roosa Master Model JDB331AL2405 injection pump; Series 1520 tractors after Serial Number 117500 are equipped with Roosa Master Model JDB331AL2406 injection pump and Series 2020 tractors after serial number 117500 are equipped with Roosa Master Model JDB431AL2408 injection pump.

Series 1530 tractors may be equipped with Roosa Master Model JDB331AL2407 injection pump and Series 2030 tractors may be equipped with Roosa Master Model JDB431AL2409 or JDB435AJ2680 injection pump.

84. **REMOVE AND REINSTALL.** To remove the injection pump, first shut off fuel and clean injection pump, lines and surrounding area. Turn engine until No. 1 piston is at TDC on compression stroke.

NOTE: Pump can be removed and reinstalled without regard to crankshaft timing position, however, TDC-1 position is necessary if timing is to be checked. If timing is not to be changed, scribe timing marks on pump flange and engine front plate which can be realigned when pump is reinstalled.

Disconnect or remove fuel inlet, return and pressure lines, throttle rod and solenoid wire from injection pump. Remove mounting stud nuts and carefully slide pump straight to rear until clear of pump shaft and seals.

The pump shaft contains two soft plastic seals which are installed back to back as shown in Fig. 68. A special tool (Roosa Master 13369) is required to install seals as shown. Only the rear seal can be installed without removing shaft from tractor (and gear). If both seals must be renewed, drain cooling

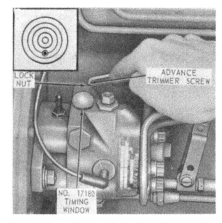

Fig. 67—Pump timing window installed to adjust intermediate advance. Concentric lines (inset) are two pump degrees apart.

Fig. 64—Mating timing marks on drive shaft and pump rotor must be aligned when pump is installed. Refer to paragraph 81.

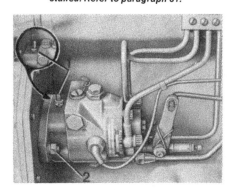

Fig. 65—Invert the threaded timing pin to properly time the pump. Pin is normally installed as shown in inset.

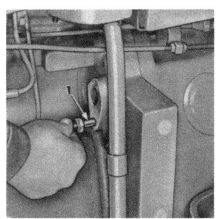

Fig. 66—Invert the threaded, flywheel timing pin (1) as shown to properly position crankshaft for injection pump installation. Refer also to Fig. 65.

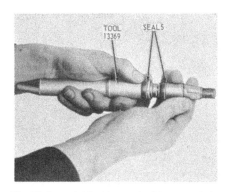

Fig. 68—A special tool is required to properly install shaft seals as shown.

system and remove lower radiator hose. Remove access plate from front of timing gear cover, back gear nut out until flush with end of shaft and jar shaft sharply with a drift and hammer to loosen gear, then withdraw shaft after removing the nut. Gear will remain in engagement with idler gear if timing gear cover is not removed.

To install new seals, first examine seal grooves in shaft carefully and remove any roughness or burrs. Coat seal liberally with Lubriplate and install from each end of shaft using the special installing tool. If shaft is removed and both seals renewed, reinstall shaft in pump before installing pump on engine. Reference mark (dot) on shaft tang and pump slot must align as shown in inset, Fig. 69.

The special Seal Installation Tool (Roosa Master 13371 or equivalent) must be used when installing pump (or shaft in pump). Also use extreme care. If resistance is felt, remove the pump and re-examine rear seal. If lip has been turned back, renew the seal.

If pump is not to be timed, realign the previously installed scribe marks and tighten stud nuts, then complete the assembly by reversing the removal procedure.

To time the pump, first be sure that crankshaft is at TDC-1. Remove timing cover from side of pump housing and turn the pump until governor weight timing line and cam timing line are in register as shown in Fig. 70. Tighten retaining cap screws to a torque of 35 ft.-lbs.

Bleed fuel system as outlined in paragraph 66 and if necessary, adjust throttle linkage as in paragraph 88.

85. **TIMING.** To check the timing without removing injection pump, turn crankshaft until No. 1 piston is coming up on compression stroke, then remove engine timing pin and cover as shown in inset, Fig. 70. Insert timing pin, long end first, into threaded hole in housing and continue turning crankshaft until

pin slides into timing hole in flywheel.

Shut off fuel and remove timing cover from side of injection pump. With crankshaft at TDC-1, the timing scribe line on governor weight retainer should align with cam timing line as shown. If it does not, loosen pump mounting stud nuts and rotate pump in slotted holes until scribe lines are aligned. Hold pump in this position and retighten mounting stud nuts.

86. ADVANCE TIMING. The injection pump is provided with automatic speed advance which is factory set and will not normally need to be checked or reset. Minor adjustments can, however, be made without removal or disassembly of the pump. To check the advance mechanism, proceed as follows:

Shut off fuel, remove pump timing hole cover and install timing window as shown in Fig. 71. Turn on fuel and bleed fuel system, then start and run engine.

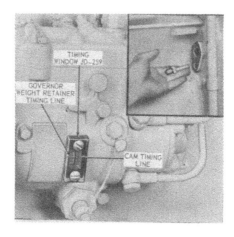

Fig. 70–With engine at TDC-1 when checked with timing pin (Inset), governor weight retainer timing line and cam timing line must register as shown. Timing window need only be installed to check intermediate advance as shown in Fig. 71.

engine speed to 1200 rpm for Series 1020 or 1520; or 1400 rpm for Series 2020; and set intermediate advance to 4° by loosening locknut and turning advance trimmer screw as shown. Tighten locknut and reinstall seal cap, then remove timing window and reinstall timing hole cover.

Roto Diesel Pump

87. **REMOVE AND REINSTALL.** To remove the injection pump, first shut off fuel and clean injection pump, lines and surrounding area. Pump can be removed and reinstalled without regard to crankshaft timing position and timing cannot be checked. The only critical requirement of the timing process is correct positioning of the injection pump gear as shown in Fig. 35.

Disconnect or remove fuel inlet, return and pressure lines, throttle rod and stop cable from injection pump. Drain radiator and remove lower radiator hose, then remove access plate from front of timing gear cover. Remove the three cap screws attaching drive gear to injection pump flange. Support pump, remove the three mounting stud nuts and pull injection pump from timing gear housing. Pump drive gear will be retained by timing gear cover.

When reinstalling pump, turn pump hub until timing slot in hub flange aligns with dowel pin in gear, and reverse removal procedure. Tighten gear mounting cap screws and flange mounting stud nuts to a torque of 18 ft.-lbs. Bleed system as outlined in paragraph 66 and adjust linkage as in paragraph 90.

SPEED AND LINKAGE ADJUSTMENT

Model DBG and JDB Pumps

88. To adjust diesel engine high and low idle speeds and the control linkage,

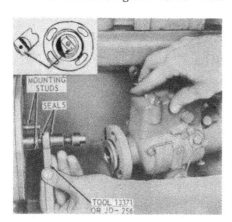

Fig. 69–Align reference marks (Inset) and use the special seal tool when installing pump.

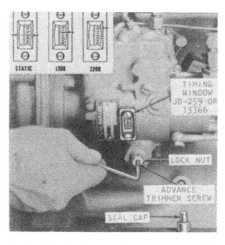

Fig. 71–Adjusting intermediate advance on JDB pump. Inset shows movement of scribe line on cam ring.

Fig. 72–Installed view of Roto-Diesel injection pump showing timing marks aligned. Pump will be in time with engine if gear timing marks are also in line.

1. Front plate
2. Pump flange
3. Engine timing mark
4. Pump timing mark

proceed as follows: Start engine and bring to operating temperature. Disconnect control rod (17—Fig. 73 or 6—Fig. 75) from injection pump, move throttle arm to high idle position and check the engine high idle speed which should be 2650 rpm for tractors without a foot throttle, or 2800 rpm for tractors with a foot throttle. If engine high idle is not as stated, break seal on pump high idle adjusting screw, turn screw as required and reseal. Now slowly move pump throttle lever to low idle position and check engine speed which should be 800 rpm. If engine low idle is not as stated, adjust pump low idle screw as required.

NOTE: Low idle position of pump throttle arm on DBG pump can be determined by the throttle arm having a small amount of free travel between the points where engine low idle is not affected and where pump goes into the fuel shut-off position.

With engine speeds adjusted at injection pump, raise cowl top door, remove right side cowl panel and adjust throttle linkage as follows: Without lifting throttle hand lever, move lever upward to the low idle position, and if necessary, adjust control rod (17—Fig. 73) until it will just enter pump throttle arm hole when pump throttle arm is positioned where it just begins to increase engine speed. Install washer and cotter pin on throttle rod. Pull throttle hand lever down toward high idle position until engine rpm reaches 2650 rpm, and if necessary,

Fig. 74–Exploded view of throttle linkage used on all early models except Low Profile.

1. Pin
2. Hub
3. Boot
4. Lever
5. Knob
6. Stop spring
7. Retaining ring
8. Lever shaft
10. Special cap screw
11. Jam nuts
12. Lever stop
14. Pin
15. Jam nuts
16. Swivel
17. Control rod
18. High idle stop screw
19. Fuel shut-off stop screw
20. Control shaft
21. Bushing
22. Arm
23. Rod clip
24. Control rod, rear
25. Pin
26. Pin
27. Shaft arm
28. Pin
29. Bushing
30. Sleeve
31. Tension nut
32. Spring
33. Lower disc
34. Facings
35. Upper disc

adjust the high idle stop screw (18) until it lacks a turn or turn and one half of contacting lever (12). If tractor is equipped with a foot throttle (38), depress against platform and check engine speed which should be 2800 rpm. If adjustment is necessary, adjust rod clevis (37) until foot pedal lacks about ⅛-inch of contacting platform. Both hand throttle and foot throttle should be adjusted to provide enough

overtravel to insure that spring on injection pump throttle arm is preloaded about 1/16-1/8-inch.

To adjust fuel shut-off position, move hand lever to low idle position, lift

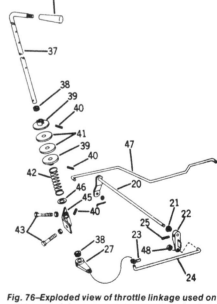

Fig. 76–Exploded view of throttle linkage used on all late models, except series 2020 Low Profile.

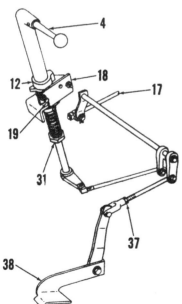

Fig. 73–Schematic view showing throttle linkage adjustments of early models, except Low Profile.

4. Hand lever	19. Fuel shut-off stop screw
12. Lever stop	31. Tension nut
17. Control rod	37. Clevis
18. High idle stop screw	38. Foot pedal

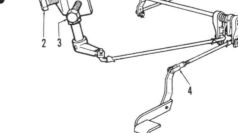

Fig. 75–Schematic view showing throttle linkage adjustments of late models.

1. Hand lever	4. Foot throttle yoke
2. Slow idle stop screw	5. Cross shaft
3. Fast idle stop screw	6. Control rod

20. Control shaft		39. Disc	
21. Bushing		40. Groove pin	
22. Control arm		41. Facings	
23. Clevis clip		42. Spring	
24. Control rod (rear)		43. Stop screws	
25. Spring pin		45. Stop	
27. Arm		46. Washer	
36. Knob		47. Control rod (front)	
37. Control lever		48. Locking cap	
38. Bushing			

lever and slowly move lever upward until engine shuts off and stop the lever travel at this point. Now turn shut-off stop screw (19) until it contacts stop (12), then back-off screw two full turns and tighten jam nut.

Move hand throttle through its range of travel to see that it will maintain any throttle setting without creeping. Movement of hand throttle should require 5 pounds pull applied at knob. Adjust by turning nut (31) as required. Reinstall cowl panel.

Model CDC and CBC Pumps

89. To adjust diesel engine high and low idle speeds and the control linkage, proceed as follows: Start engine and run until it reaches operating temperature.

On Model CDC pump, refer to Fig. 77 and remove slow idle adjusting screw from end of throttle cap. Disconnect control rod from injection pump throttle lever, move pump throttle lever to high idle position and check the engine high idle speed which should be 2630-2670 (2650 desired) rpm for tractors without a foot throttle, or 2780-2820 (2800 desired) rpm for tractors with a foot throttle. If engine high idle is not as stated, insert a small bladed screw driver in rear of throttle control cap and turn high idle adjusting screw in to increase, or out to decrease, the engine high idle speed.

Reinstall the low idle adjusting screw, move injection pump throttle lever to low idle position and adjust low idle screw to obtain an engine low idle speed of 800 rpm. Tighten low idle adjusting screw lock nut.

NOTE: After low idle adjustment is made be sure that at least ⅛-inch clearance remains between pump housing and throttle lever when lever is at low idle position. If necessary, loosen the lever retaining screws and reposition throttle lever.

On Model CBC pump, disconnect control rod from injection pump throttle arm (7—Fig. 78). Pull throttle

arm away from injection pump to high idle stop and check engine speed which should be 2800 rpm. To adjust high idle speed, remove throttle cap (3) and turn adjusting nut (4) until specified high idle speed is attained. Reinstall throttle cap (3), move throttle arm toward injection pump and check slow idle speed which should be 650 rpm. To adjust slow idle speed, loosen jam nut (2) and turn stop screw (1) to specified setting. With slow idle speed adjusted, check to be sure lever arm (7) clears pump body (8) by approximately ⅛-¼ inch. If it does not, loosen the two clamp screws (6) and reposition lever arm until correct measurement is attained.

With engine speeds adjusted at injection pump, reconnect control rod to pump throttle lever. Raise cowl top door, remove right side cowl panel and adjust linkage as follows: Locate the stop screws (2 & 3—Fig. 75) on lower end of throttle hand lever and turn low idle screw in several turns. Move throttle hand lever to idle position, then continue to push hand lever upward until upper end of injection pump throttle lever has about ¼-inch over-travel. Keep hand throttle lever in this position, back out the low idle adjusting screw until it contacts stop and tighten jam nut.

If tractor does not have a foot throttle, turn high idle screw in several turns, then move hand throttle lever to high idle position. Continue to pull hand lever down until top of injection pump throttle lever has about ¼-inch over-travel. Keep hand lever in this position, back out the high idle screw until it contacts stop and tighten jam nut.

If tractor is equipped with a foot throttle, adjust linkage as follows: Depress foot throttle against platform and adjust foot pedal rod length until upper end of injection pump throttle lever has about ¼-inch over-travel. Tighten the clevis jam nut. Place hand throttle lever in high idle position and adjust idle screw to obtain the recommended high idle speed of 2650 rpm, then tighten jam nut.

Roto Diesel Pump

90. To adjust diesel engine speeds and control linkage, start engine and run until operating temperature is reached. Disconnect speed control rod from injection pump and move throttle lever (3—Fig. 79) against fast idle adjusting screw (2). Engine speed should be 2650 rpm. If it is not, turn fast idle screw (2) in or out as required. Move pump throttle lever against slow idle adjusting screw (1); engine speed should be 650 rpm. If it is not, adjust slow idle screw. Reconnect throttle control rod (6—Fig. 75) to pump and adjust stop screws (2 & 3) as required until pump throttle lever (3—Fig. 79) contacts fast and slow stop screws (1 & 2). If foot throttle is used, adjust foot throttle linkage until pedal pad contacts footrest at the same time pump throttle lever (3) contacts fast idle screw (2).

To adjust shut-off cable, completely push in stop knob and check to be sure stop lever contacts stop on injection pump governor cover. If it does not, loosen cable clamp screw and reposition clamp on stop cable.

INJECTION PUMP SHUT-OFF

Models CDC, CBC, JDB

91. These models are equipped with a solenoid type shut-off which must be energized to provide fuel flow. The shut-off valve is spring loaded in the closed position and is opened when key

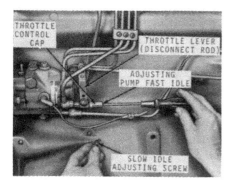

Fig. 77—Remove low idle adjusting screw and insert screw driver in end of throttle control cap to adjust the high idle screw on Model CDC pump.

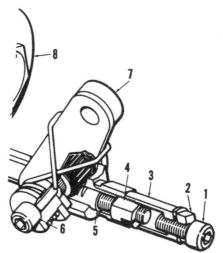

Fig. 78—Cross sectional view of governor adjustment on Model CBC injection pump; refer to paragraph 89.

1. Slow idle stop	
2. Jam nut	
3. Throttle cap	5. Governor rod
4. High speed adjusting	6. Clamp screw
nut	7. Lever arm
	8. Pump body

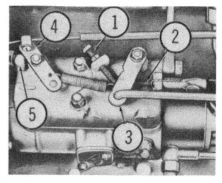

Fig. 79—Installed view of Roto-Diesel pump showing linkage adjustments.

1. Slow idle screw	
2. Fast idle screw	4. Stop lever
3. Pump throttle arm	5. Lever stop

switch is turned to "ON" or "START" position.

If tractor will not start, turn switch to "ON" and check continuity of solenoid lead.

GASOLINE ENGINE GOVERNOR

All Gasoline Models

The governor for gasoline engines is a centrifugal flyweight type which is mounted to the engine front plate on left side of engine. The governor is gear driven from the upper idler of the engine gear train. The governor shaft forward end is supported by a bushing in the engine timing gear cover.

92. LINKAGE ADJUSTMENT. To adjust linkage, use Figs. 73 through 76 as a reference, except identify item number (19—Fig. 73) as Low Idle Stop Screw instead of Fuel Shut off Stop Screw.

Adjust linkage as follows. Disconnect the governor to carburetor rod at governor, hold both governor and carburetor in the wide open position and

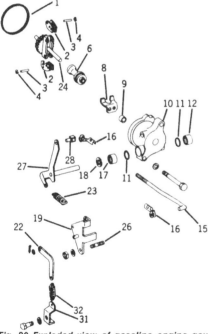

Fig. 80–Exploded view of gasoline engine governor unit. Use sealant on outside of closed end bearing (12) during assembly.

1.	Seal	16.	Rod clip
2.	Weights	17.	Bearing
3.	Weight pins	18.	Oil seal
4.	Retainers	19.	Speed change lever
6.	Sleeve and thrust	22.	Counterbalance arm
	bearing	23.	Governor spring
8.	Fork	24.	Shaft and gear
9.	Bushing	26.	Stud
10.	Housing	27.	Governor lever
11.	Thrust washer	28.	Swivel
12.	Bearing (closed end)	31.	Bracket
15.	Throttle rod	32.	Spring

adjust rod length until swivel will just fit into hole in governor arm, then shorten the rod one full turn of the swivel. Lift cowl top door, remove cowl right side panel and pre-set low idle stop screw (19) so head protrudes 7⁄8-inch from support bracket. Tighten jam nut. Start engine and bring to operating temperature, then without lifting hand lever, move throttle to low idle position. Check engine speed which should be 600 rpm and if adjustment is necessary, turn the carburetor throttle stop screw as required. Disconnect control rod at rear end and on tractors without foot throttle, hold rod forward so carburetor throttle stop screw is against its stop. Be sure hand throttle is in low idle position, then lengthen control rod (17) one full turn. On models with foot throttle, use same procedure and adjust control rod (17) so that outer arm on cross-shaft moves 1/16-inch away from inner arm. Reconnect control rod, place hand throttle lever in high idle position and check engine speed which should be 2680 rpm. If engine speed is not as stated, adjust high idle stop screw (18) as required. If tractor is equipped with a foot throttle, depress pedal until it contacts platform and check engine speed which should be 2800 rpm. If engine speed is not correct, adjust clevis (37) as required.

Move hand throttle through its range of travel to see that it will maintain any throttle setting without creeping. Movement of hand throttle should require 10 pounds pull applied at knob. Adjust by turning nut (31) as required. Reinstall cowl panel.

93. R&R AND OVERHAUL. To remove governor, first remove fan baffle (if so equipped) and carburetor inlet tube. Disconnect throttle rod and control rod at governor and unhook spring from counterbalance arm. Re-

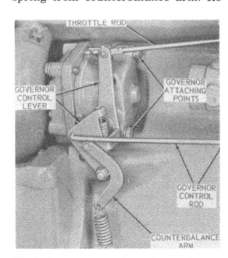

Fig. 81–Installed view of gasoline engine governor showing component parts of linkage.

move lock nut from speed change lever stud, unhook governor spring and remove the speed change lever and counterbalance arm. Remove governor retaining cap screws and pull governor assembly from engine front plate.

With governor removed, pull weight carrier assembly from housing and remove seal (1—Fig. 80) from housing front flange. Remove fork (8) from lever shaft (27) and pull shaft from housing. Any further disassembly required is obvious and will be dictated by the need of parts renewal. A new housing (10) is fitted with items (8, 9, 11, 12, 17, 18 and 27).

When installing bearings (12 and 17), press on lettered end of bearing (17) and on closed end of bearing (12) and press both bearings into housing until inner ends are flush with inside of housing. Coat outside of bearing (12) with sealing compound prior to installation. Coat lip of seal (18) with Lubriplate and install with lip facing inward. Bushing (9) is pressed into its bore until bottomed. Weights (2) are a free fit on weight pins (3), however weight pins must be tight fit in weight carrier (24).

Reassemble by reversing disassembly procedure and after governor unit is installed, adjust linkage as outlined in paragraph 92.

COOLING SYSTEM

RADIATOR

All Models

94. Some models have an oil cooler attached to right side of radiator as shown in Fig. 82. Except for discon-

Fig. 82–Hydraulic oil cooler (OC) is located at side of coolant radiator as shown.

necting hoses from the oil cooler, removal procedure for all models is similar.

95. REMOVE AND REINSTALL. To remove radiator, first drain cooling system, then remove grille screens and hood. Remove air intake tube. Remove fan shroud from radiator and lay shroud back over fan. On diesel models, disconnect injector leak-off line from fuel tank. On models so equipped, disconnect inlet line from top of oil cooler, the two small bleed lines from top of hydraulic reservoir and the outlet line from bottom of oil cooler. On all models, disconnect upper and lower radiator hoses and the upper radiator brace from radiator, then unbolt and remove radiator from tractor.

WATER PUMP

All Models

96. REMOVE AND REINSTALL. To remove water pump, first remove radiator as outlined in paragraph 95, then remove fan and fan belt. Disconnect by-pass hose from water pump, then unbolt and remove water pump from engine.

Reinstall by reversing removal procedure and adjust fan belt so a 25 lb. pull mid way between pulleys will deflect belt ⅝-inch.

97. OVERHAUL. To disassemble water pump, use Fig. 83 as a guide and proceed as follows:

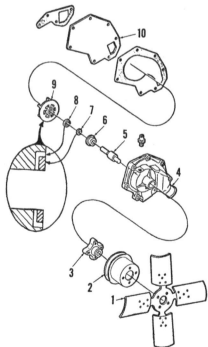

Fig. 83–Exploded view of water pump showing component parts.

1. Fan blades	6. Seal
2. Pulley	7. Insert
3. Hub	8. Cup
4. Body	9. Impeller
5. Shaft & bearing	10. Cover

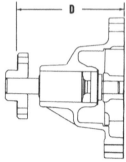

Fig. 84–Distance (D) from front face of pulley hub to gasket surface of body should be approximately 5½ inches when pump is properly assembled.

Support fan pulley hub in a press and, using a suitable mandrel, press shaft from pulley hub. Suitably support housing on gasket surface and press shaft, bearing, seal and impeller as a unit from housing.

Bearing outer race is a tight press fit in housing bore and is not otherwise secured. It is important therefore, that reasonable precautions be taken during assembly to prevent bearing movement during installation of impeller and pulley hub.

Coat outer edge of seal (6) with sealant and install in housing (4) using

a socket or other driver which contacts only the outer flange of seal, then insert long end of bearing (5) through front of housing bore. Use tool No. JD-262 or a similar tool which contacts only outer race of bearing and press bearing into housing until front of bearing is flush with housing bore.

Install insert (7) in cup (8) with "V" groove of insert toward cup. Parts must be clean and dry. Dip cup and insert in engine oil then press cup and insert into impeller as shown in inset, until cup bottoms in impeller counterbore. Support front end of pump shaft and press impeller (9) on rear of shaft until fins are flush with gasket surface of housing.

Invert the pump assembly and support the unit on rear of shaft which is recessed into impeller hub. DO NOT support impeller or housing. With shaft suitably supported, press pulley hub (3) on front of shaft until front face of pulley hub is approximately 5½ inches from gasket surface of housing as shown at (D—Fig. 84). NOTE: Front of shaft will be approximately flush with hub bore as shown. Complete the assembly by reversing the disassembly procedure.

IGNITION AND ELECTRICAL SYSTEM

DISTRIBUTOR

All Gasoline Models

98. TIMING. A "TDC" timing hole is located in flywheel, but ignition is timed with an "S" or "28" mark on crankshaft pulley which aligns with a scribed line on right side of timing gear cover when advance spark should occur. Refer to Fig. 85. Suggested timing method is by using a power timing light and timing with marks aligned at 2500 engine rpm.

Recommended breaker point gap is 0.020 and distributor maximum advance is 30 crankshaft degrees.

99. REMOVE AND REINSTALL. Before removing the distributor assembly, turn crankshaft until No. 1 piston is at TDC on compression stroke. Remove distributor cap and disconnect coil lead. Remove distributor clamp screw then withdraw distributor assembly from cylinder block. When installing distributor, refer to Fig. 86 and proceed as follows:

Turn distributor housing until primary lead is pointing directly toward engine, then turn distributor shaft until rotor arm is pointing about 75 degrees clockwise from primary lead as shown. With No. 1 piston at TDC on

compression stroke, install the distributor. With gears fully meshed, rotor should point about 50 degrees clockwise from primary lead. Complete the assembly, start engine and time with a timing light as outlined in paragraph 98.

NOTE: An error of one tooth in installation timing will make about 30° difference in rotor position and is readily apparent after referring to Fig. 86.

Firing order is 1-2-3 for three cylinder models and 1-3-4-2 for four cylinder

Fig. 85–Distributor is correctly timed when "S" or "28" mark on crankshaft pulley aligns with timing mark on engine casting in advance timing position.

inder models. Recommended distributor point gap is 0.020.

100. **OVERHAUL.** Prestolite distributors are used. Breaker point gap is 0.020 and breaker point spring tension should be 17-22 oz. measured at center of contact. Tension can be adjusted by sliding the spring in or out on attaching screw. Direction of rotation is counter-clockwise, viewed from rotor end of shaft. Distributor shaft end play should be 0.002-0.010, with a specified wear limit of 0.015.

Distributor shaft bushings may be renewed in housing and both bushings should be installed 3/32-inch below flush with housing bore. Grease hole must be drilled in upper bushing after installation, after removing grease hole plug and wick. Deburr and clean the bushing and relubricate the wick with high melting point grease.

When distributor is assembled at factory, the spring pin (roll pin) which retains drive gear is left protruding about 1/16-inch on the end which aligns with rotor arm.

Cam dwell angle is 96-102 degrees for three cylinder models or 66-72 degrees for four cylinder units. Advance data follows, given in distributor degrees and distributor rpm. Intermediate advance can be adjusted if necessary, by bending outer spring posts.

Three Cylinder Models
Start Advance .... 0° @ 200 rpm
Intermediate advance:
  IBT 4301B or C ...... 3° @ 350 rpm
  IBT 4301D .... 3° @ 400 rpm
Maximum advance ... 15° @ 1200 rpm
Four Cylinder Models
Start advance .... 0° @ 200 rpm
Intermediate advance .. 4° @ 400 rpm
Maximum advance ... 15° @ 1200 rpm

ALTERNATOR AND REGULATOR

Motorola Alternator

101. Refer to Fig. 87 for an exploded view of MOTOROLA alternator unit.

The isolation diode and brush holder

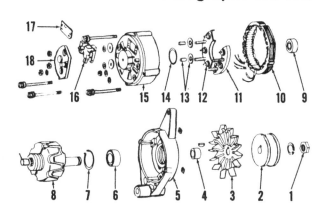

Fig. 87–Exploded view of MOTOROLA alternator unit showing component parts.

1. Nut
2. Pulley
3. Fan
4. Spacer
5. Drive end frame
6. Bearing
7. Snap ring
8. Rotor
9. Bearing
10. Stator
11. Negative heat sink
12. Positive heat sink
13. Insulators
14. Retainer
15. Slip ring end frame
16. Brush holder
17. Cover
18. Isolation diode

can be renewed without removal or disassembly of alternator; all other alternator service requires disassembly.

The primary purpose of the isolation diode is to permit use of charging indicator lamp. Failure of the isolation diode is usually indicated by the indicator light; which glows with engine stopped and key switch off if diode is shorted; or with engine running if diode is open.

Failure of a rectifying diode may be indicated by a humming noise when engine is running, if diode is shorted; or by a steady flicker of charge indicator light at slow idle speed if diode is open. Either fault will reduce alternator output.

To check the charging system, refer to Fig. 88 and proceed as follows:

(1). With key switch and all accessories off and engine not running, connect a low reading voltmeter to terminals D-F. Reading should be 0.1 volt or less. A higher reading would indicate a short in isolation diode, key switch or wiring.

(2). Turn key switch on but do not start engine. Recheck voltmeter reading which should be 1-3 volts. A higher or lower reading may indicate a defective alternator, regulator or wiring.

(3). Start and run engine at approximately 1400 rpm and, with all accesso-

ries off, again check voltmeter reading which should be 15 volts. A lower reading could indicate a discharged battery or defective alternator.

(4). Move voltmeter lead from auxiliary terminal (D) to output terminal (E) and recheck voltage. Reading should drop one volt from reading in previous test (3), reflecting the resistance designed into isolation diode (18—Fig. 87). If battery voltage (12 volts) is obtained, isolation diode is open and must be renewed.

(5). If a reading lower than the specified 15 volts was obtained when checked as outlined in test 3, stop engine and disconnect regulator plug (B). Connect a jumper wire between output terminal (E) and field terminal (C) on alternator brush holder. Connect a suitable voltmeter to terminals (D-F) on alternator. Start engine and slowly increase engine speed while watching voltmeter. If a reading of 15 volts can now be obtained at 1300 engine rpm or less, renew the regulator. If a reading of 15 volts cannot be obtained, renew or overhaul the alternator.

CAUTION: DO NOT allow voltage to rise above 16.5 volts when making this test. DO NOT run engine faster than 1500 rpm with regulator disconnected.

102. **OVERHAUL.** The isolation diode and brush holder can be removed without removing alternator from

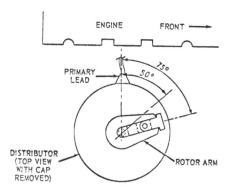

Fig. 86–Rotor arm should be positioned approximately 75 degrees from primary lead before distributor is installed. With No. 1 piston at TDC, rotor should point about 50 degrees from lead with distributor installed.

Fig. 88–Wiring diagram of charging system showing test locations. Refer to paragraph 101 for test procedure.

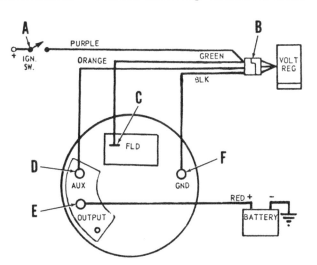

tractor. Brush holder should be removed before attempting to separate the frame units.

To disassemble the removed alternator unit, remove through-bolts and separate slip ring end frame (15—Fig. 87) from drive end frame (5). Rotor (8) will remain with drive end frame and stator (10) with slip ring end frame. Be careful not to damage stator windings when prying units apart.

Examine slip ring surfaces of rotor for scoring or wear and field windings for overheating or other damage. Check bearing surfaces of rotor shaft for visible wear or scoring. Check rotor for grounded, shorted or open circuits using an ohmmeter as follows:

Refer to Fig. 89 and touch the ohmmeter probes to points (1-2) and (1-3); a reading near zero will indicate a ground. Touch ohmmeter probes to the two slip rings (2-3); reading should be 5.5 ohms. A higher reading will indicate an open field circuit, a lower reading a short.

Runout should not exceed 0.002. Slip ring surfaces can be trued if runout is excessive or if surfaces are scored. Finish with 400 grit or finer polishing cloth until scratches or machine marks are removed.

Stator is "Y" connected (Fig. 90) and center connection need not be unsoldered to test for continuity. Each field winding uses two coils as shown. Continuity should exist between any two of the stator leads but not between any lead and stator frame. Because of the low resistance, shorted windings

within a coil cannot be satisfactorily checked. Three positive diodes are located in slip ring heat sink (12—Fig. 87) and three negative diodes in grounded heat sink (11). Diodes should test at or near infinity in one direction when checked with an ohmmeter, and at or near zero when meter leads are reversed. Renew any diode with approximately equal meter readings in both directions. Diodes must be removed and installed using an arbor press or vise and a suitable tool which contacts only outer edge of diode. Do not attempt to drive a faulty diode out of heat sink, as shock may cause damage to other good diodes. If all diodes are being renewed, make certain the positive diodes (marked with red printing) are installed in positive heat sink (12) and negative diodes

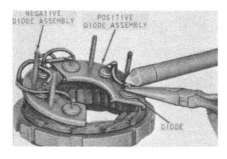

Fig. 91—Use needle nose pliers as a heat sink, and an iron only, when soldering diode connections.

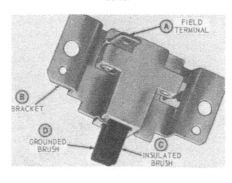

Fig. 92—Check brush holder for continuity between A-C; and B-D.

(marked with black printing) are installed in negative heat sink (11). Use a pair of needle nose pliers as a heat sink when soldering diode leads (See Fig. 91). Use only rosin core solder and an iron instead of a torch. Excess heat can damage a good diode while it is being installed.

Exposed length of brushes in removed brush holder should be ¼-inch or more. Brushes are available only in an assembly with the holder. Check for continuity between field terminal ((A—Fig. 92) and insulated brush (C); and between brush holder (B) and grounded brush (D). Wiggle the brush and lead while checking, to test for poor connections or an intermittent ground.

NOTE: A battery powered test light can be used instead of an ohmmeter for all electrical tests except shorts in rotor winding. When checking diodes however, test light must not be more than 12 volts.

Bosch Alternator

103. **OPERATION AND TESTING.** Refer to Fig. 93 for exploded view of BOSCH 12 Volt, 28 Ampere alternator used on some models.

To check the charging system, connect a voltmeter to B+ (Output) terminal (1—Fig. 94) on alternator and to

Fig. 94—Installed view of Bosch alternator.

1. Output terminal
2. Terminal plug
3. D+ terminal

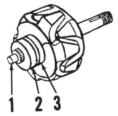

Fig. 89—Removed rotor assembly showing test points to be used when checking for grounds, shorts and opens.

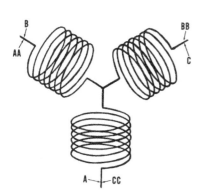

Fig. 90—Schematic view of typical "Y" connected stator. Center connection need not be unsoldered to check for continuity.

Fig. 93—Exploded view of Bosch alternator used on some models.

1. Brush holder
2. Rectifier
3. Stator
4. Rotor

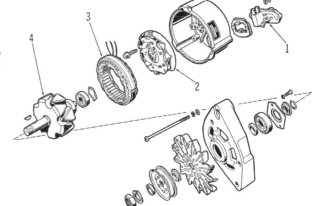

a suitable ground. With engine running at 1200 rpm, reading should be 13 volts or above. A lower reading could indicate a discharged battery or faulty alternator or regulator.

With engine not running, disconnect the three-terminal plug (2) from alternator and touch ammeter leads to DF (Green Wire) terminal and B+ terminal. Current draw should be approximately 2 Amperes. High readings are caused by shorts or grounds. Low readings may be caused by dirty slip rings or defective brushes.

Connect a jumper wire between B+ and DF terminals and a voltmeter between B+ terminal and ground. Start engine and increase engine speed until voltage rises not to exceed 15.5 volts. If maximum reading is less than 14 volts and battery is fully charged, alternator is defective.

Move jumper wire connection from B+ terminal to D+ terminal. Voltage reading should remain the same. If voltage drops, exciter diodes are defective. If voltage reading was normal with plug disconnected and jumper wire installed, but was low when tested with plug connected, renew the regulator.

104. **OVERHAUL.** Brush holder (1 —Fig. 93) can be removed without removing alternator from the tractor. Brush holder should be removed before attempting to separate alternator frame units.

To disassemble the removed alternator, first remove capacitor and brush holder. Immobilize pulley and remove shaft nut, pulley and fan. Mark brush end housing, stator frame and drive

end housing for correct reassembly and remove through-bolts. Rotor will remain with drive-end frame and stator will remain with brush-end frame when alternator is disassembled.

Remove the two terminal nuts and three screws securing rectifier (2) to brush end frame and lift out rectifier and stator (3) as a unit. Carefully tag the three stator leads for correct reassembly, then unsolder the leads from rectifier diodes using an electric soldering iron and minimum heat. Be careful not to get solder on diode plates or overheat the diodes.

Check brush contact surface of slip rings for burning, scoring or varnish coating. Surfaces must be true to within 0.002 inch. Contact surface may be trued by chucking in a lathe. Polish the contact surface after truing using 400 grit polishing cloth, until scratches and machine marks are removed. Check continuity of rotor windings using an ohmmeter as shown in Fig. 96. Ohmmeter reading should be 4.0— 4.4 between the two slip rings and infinity between either slip ring and rotor pole or shaft.

Stator is "Y" wound, the three individual windings being joined in the middle. Test the windings using an ohmmeter as shown in Fig. 97. Ohmmeter reading should be 0.4-0.44 between any two leads and infinity between any lead and stator frame.

Alternator brushes and shaft bearings are designed for 2000 hours ser-

vice life. New brushes protrude 0.4 inch beyond brush holder when unit is removed; for maximum service reliability, renew BOTH the brushes and shaft bearings when brushes are worn to within 0.2 inch of holder. Solder copper leads to allow for 0.4 inch protrusion using rosin core solder only, and making sure the solder does not seep into and stiffen the wire lead.

The rectifier is furnished as a complete assembly and diodes are not serviced separately. The rectifier unit contains three positive diodes, three negative diodes and three exciter diodes which energize rotor coils before engine is started. If any of the diodes fail, rectifier must be renewed.

To test the positive diodes, touch positive ohmmeter probe to positive heat sink as shown in Fig. 98, and touch negative test probe to each diode lead in turn. Ohmmeter should read at or near infinity for each test. Reverse the leads and repeat the series; ohmmeter should read at or near zero for the series.

Test negative diodes as shown in Fig. 99. Place negative test probe on negative heat sink and touch each diode lead in turn with positive test probe. Ohmmeter should read at or near infinity for the series. Reverse test leads and repeat the test; ohmmeter should read at or near zero for the series.

Test exciter diodes by using the D+ terminal as the base as shown in Fig. 100. Ohmmeter should read at or near

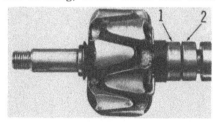

Fig. 95–A reading of 4.0-4.4 Ohms should exist between slip rings (1 & 2) when tested with an ohmmeter.

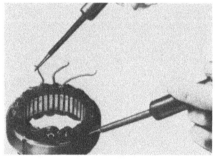

Fig. 97–No continuity should exist between any stator lead and stator frame.

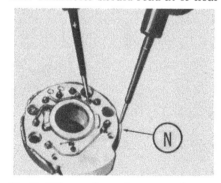

Fig. 99–With negative probe on negative heat sink (N) and positive probe touching diode leads, an infinity reading should be obtained. Reverse the probes and reading should be near zero.

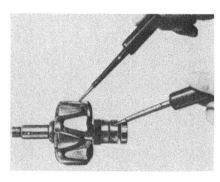

Fig. 96–No continuity should exist between either slip ring and any part of rotor frame.

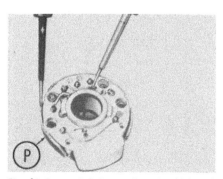

Fig. 98–A near-infinity reading should be obtained when positive probe rests on positive heat sink (P) and negative probe touches diode leads as shown. Reverse the probes and reading should be near zero Ohms.

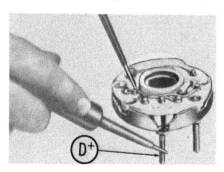

Fig. 100–D+ terminal is used to test exciter diodes, refer to paragraph 104.

infinity with positive test probe on terminal screw and at or near zero with negative test probe touching screw.

When assembling alternator, tighten through-bolts to approximately 35-40 inch-pounds and pulley nut to 25-30 ft.-lbs.

STARTING MOTOR

All Models

105. Delco Remy or Bosch starting motors are used. Test specifications, including solenoid, are as follows:

D-R 1107339
Brush spring tension .... 35 oz.
No load test:
    Volts 10.6
    Min. amps 65
    Max. amps 100
    Min. rpm 3600
    Max. rpm 5100

D-R 1107577
Brush spring tension .... 35 oz.
No load test:
    Volts 10.6
    Min. amps 105
    Max. amps 200
    Min. rpm 6500
    Max. rpm 14000

D-R 1107599
Brush spring tension .... 35 oz.
No load test:
    Volts 10.6
    Min. amps 105
    Max. amps 200
    Min. rpm 6500
    Max. rpm 14000

D-R 1107863
Brush spring tension .... 35 oz.
No load test:
    Volts 10.6
    Min. amps 105
    Max. amps 200
    Min. rpm 6500
    Max. rpm 14000

D-R 1108319
Brush spring tension .... 35 oz.
No load test:
    Volts 9
    Min. amps 55
    Max. amps 80
    Min. rpm 3500
    Max. rpm 6000

R. Bosch 0001 359 016
Brush spring tension .... 35-45 oz.
No load test:
    Volts 11.5
    Min. amps 90
    Max. amps 110
    Min. rpm 6000
    Max. rpm 8000

CIRCUIT DESCRIPTION

All Models

106. All models are equipped with a 12 volt electrical system with the negative battery post grounded. Diesel

models use two batteries connected in parallel.

Fig. 101 shows wiring connections to key switch. Fuel gage sender unit has a full tank resistance of 30 ohms and empty tank resistance of 0 ohms, with an even rate of increase as float lever is raised. A good ground must exist for both the sender and receiver units. Oil pressure indicator lamp switch should close at 5.5-10.5 psi.

Injection pump solenoid winding should have a current draw of approximately 2.5 amperes and a resistance of approximately 5 ohms. Carburetor shut-off solenoid winding should have a current draw of approximately 0.6 amperes and a resistance of approximately 20 ohms.

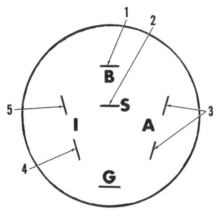

Fig. 101–Key switch viewed from terminal side, showing wiring connections.

1. Red
2. Black
3. Purple
4. Resistance wire
5. Gray

ENGINE CLUTCH

ADJUSTMENT

Series 2020 Low Profile

107. **CLUTCH ADJUST.** To check and adjust clutch, disengage clutch lever (fully forward) and remove the large button plug from left side of clutch housing. See Fig. 102. Measure the clearance between clutch throw-out bearing and clutch fingers which should be 0.090. If clearance is not correct loosen the locking screw, then turn

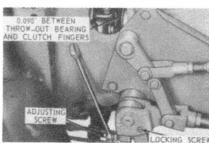

Fig. 102–Clearance between throw-out bearing and clutch finger should be 0.090.

adjusting screw as required. Tighten locking screw and jam nut on adjusting screw and install button plug.

108. **LINKAGE ADJUST.** If linkage has been disconnected, or disassembled, adjust linkage as follows: Back out throw-out bearing adjusting screw until rounded end is flush with lever. Back out locking screw, then adjust clutch lever stop screw, shown in Fig. 103, until clutch lever is in a straight up and down position. Engage clutch lever, then position a straightedge between centers of spring yoke pin and clutch arm pin and measure distance between straightedge and center of upper link pin. (See Fig. 104). This distance should be ½-⅝-inch and if adjustment is necessary, disconnect clutch lever rod and rotate clevis as required.

Engage and disengage clutch and if effort required (either direction) is excessive, check the length of clutch spring which should be about 8¼ inches as shown in Fig. 105.

After clutch linkage is adjusted, readjust throwout bearing as outlined in paragraph 107.

All Other Models

109. **FREE PLAY AND PEDAL POSITION.** Clutch pedal free play

Fig. 103–Adjust clutch lever stop screw to position clutch lever to a vertical position.

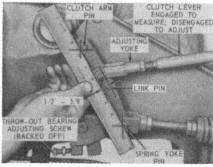

Fig. 104–Clutch linkage is correctly adjusted when above conditions are met. Refer to text.

Fig. 105–Adjust clutch spring to a length of 8¼ inches to obtain correct clutch lever operating effort.

should be one inch and should be readjusted when free play decreases to ½-inch. Adjustment is made by disconnecting yoke (Fig. 106) from clutch pedal arm and shortening or lengthening clutch operating rod as required.

Clutch pedal is used in two positions. When tractor has no pto, independent pto, or transmission driven pto, loosen pedal positioning cap screw (Fig. 107) and move clutch pedal forward until screw contacts rear of slot. Tighten cap screw securely to maintain proper pedal position. On models with continuous pto and dual clutch, move pedal pad to rear until positioning cap screw contacts front of slot; to allow clutch to fully release.

110. **RELEASE LEVERS (DUAL CLUTCH).** The dual clutch fingers can be adjusted for wear without disassembly or removal of unit from tractor. Proceed as follows:

Disconnect clutch operating rod from clutch pedal arm (Fig. 106) and refer to Fig. 109. Remove access cover from clutch housing and back off clutch operating bolt nuts until operating lever contacts pto clutch plate pins for all three operating levers. Tighten one jam nut until finger begins to pull away from pto clutch plate pin, tighten nut an additional 2½ turns and secure by tightening locknut. Adjust clutch operating rod (Fig. 106) until, with yoke pin inserted, throwout bearing just contacts operating lever. Turn flywheel until each of the other clutch levers are in position and adjust the lever to lightly contact throwout bearing. Tighten all locknuts securely and adjust clutch pedal free travel as outlined in paragraph 109.

TRACTOR SPLIT

All Models

111. To split engine from clutch housing, proceed as follows: Drain

cooling system and remove hood. Drain transmission case. Disconnect battery cables, remove battery (or batteries) and the wood (insulator) block, then remove the two cap screws which retain cowl to flywheel housing. Disconnect wiring from starter solenoid, alternator and on gasoline tractors, the ignition coil and move wiring harness rearward. Disconnect return lines from hydraulic reservoir and loosen line clip at right rear of cylinder head. Disconnect pressure line from right side of power steering valve housing and return line from nipple of hydraulic lift return line. Disconnect main hydraulic pump pressure line at coupling near front of tractor. Remove line retainer from lower right front side of transmission case and loosen inlet and return lines from housing. Do not lose check valve located in main pump return line. Disconnect temperature sending bulb from water outlet elbow and cold weather starting aid line from inlet manifold. Disconnect wire from oil pressure sending unit. Disconnect tachometer cable at clutch housing. Disconnect throttle control rod from injection pump, or governor on gasoline tractors. Disconnect choke cable from

carburetor on gasoline tractors. Disconnect drag link at either end. Place floor jack under transmission, attach hoist to front section of tractor, then unbolt engine from clutch housing and split (separate) tractor.

When rejoining tractor it may be necessary to bar over engine to facilitate entry of input shafts into clutch discs. Be sure flywheel housing and clutch housing are butted together before tightening retaining cap screws. Tighten retaining cap screws to 170 ft.-lbs. torque.

R&R AND OVERHAUL

Series 1530—Mannheim Built 2030 With Independent PTO

112. Refer to Fig. 110 for an exploded

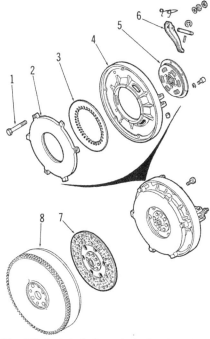

Fig. 109—Wear adjustment of dual clutch levers can be made as shown. Refer to paragraph 110.

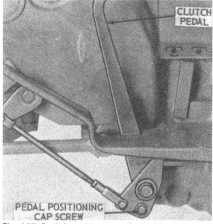

Fig. 107—Pedal position is adjusted by loosening cap screw and shifting pedal in slotted hole.

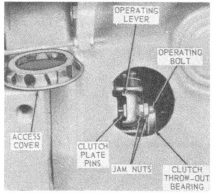

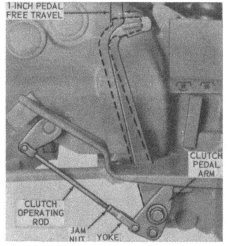

Fig. 106—External clutch adjustment is made by disconnecting rod yoke from pedal arm and adjusting length of operating rod.

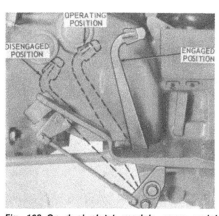

Fig. 108—On dual clutch models, move pedal rearward until cap screw contacts front of slot (Fig. 107) to permit full disengagement of pto clutch.

Fig. 110—Exploded view of single disc clutch used on late German built tractors with IPTO.

1. Through-bolt	5. IPTO drive damper
2. Pressure plate	6. Release lever
3. Diaphragm spring	7. Clutch disc
4. Cover	8. Flywheel

view of clutch unit and to Fig. 111 for cross sectional view. Clutch can be removed after clutch split as outlined in paragraph 111. When installing clutch disc, make sure long hub is forward as shown in Fig. 111. Use a suitable alignment tool and tighten cover retaining cap screws to a torque of 35-40 ft.-lbs.

To disassemble the clutch cover, back off the adjusting nuts until pressure of the diaphragm washer (spring) (3—Fig. 110) is free. Mark pressure plate (2) and cover (4) with paint so balance can be maintained, remove the nuts and separate plate and cover. Examine splines and damper springs in pto damper (5) and renew if damaged. Check diaphragm spring (3) for heat checks, cracks or distortion and renew if questionable. Check release levers and pivots for wear. Renew clutch disc (7) if facing wear approaches rivet heads, if hub is loose, or if splines are worn.

Assemble by reversing the disassembly procedure. Adjust release levers after installation as shown in Fig. 112, until lever height is 1.791-1.811, measured from torsion damper hub. Adjust linkage after tractor is reconnected, as outlined in paragraph 109.

Auburn Type Single Clutch

113. Refer to Fig. 113 for exploded view of Auburn Type single stage clutch used on Series 1520, 2020 or 2030 without pto or with transmission pto. Series 1020 is similar. Fig. 115 shows clutch cover and associated parts used on Series 1520 and 2020 with independent pto. A cross sectional view is shown in Fig. 114.

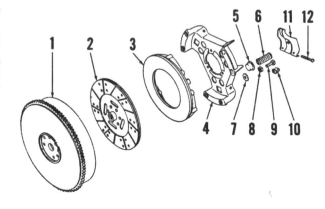

Fig. 113–Exploded view of Auburn type single disc clutch unit used on some tractors without pto or with transmission driven pto unit. Refer to Fig. 115 for view of clutch cover used on some models with independent pto.

1. Flywheel
2. Clutch disc
3. Pressure plate
4. Cover
5. Spring cup
6. Spring
7. Stop washer
8. Locknut
9. Adjusting screw
10. Return clip
11. Release lever
12. Pivot pin

Clutch can be removed after clutch split as outlined in paragraph 111. When installing clutch, make sure long hub faces forward as shown in Fig. 114. Use a suitable alignment tool and tighten clutch cover retaining cap screws to a torque of 35 ft.-lbs.

To disassemble the clutch cover assembly, support cover at mounting pads. Use a steel plate to apply pressure evenly to all three release levers. Depress the levers until enough clearance exists beneath pressure plate to allow removal of adjusting screws (9—Fig. 113), loosen locknuts and remove screws, stop washers (7) and return clips (10). Slowly release the pressure, swing back the levers and remove clutch springs (6) and if so equipped, spring cups (5).

Several different clutch springs have been used; refer to the following chart for test data and application.

Model & Ser. No.	Color Code	Minimum Test @ Ins.
1020 (Trans-Mission PTO) 0-60686	White Stripe	100 lbs. @ 1½ In.
60687-117500	Red Stripe	148 lbs. @ 1½ In.
117500	Light Blue	158 lbs. @ 1 11/16 In.
1520 (Transmission PTO) All	Light Blue	158 lbs. @ 1 11/16 In.
2020 (Transmission PTO) 0-58219	White	141 lbs. @ 1 11/16 In.
58220	Light Blue	158 lbs. @ 1 11/16 In.
All Models, Independent PTO All	Green Stripe	161 lbs. @ 1 9/16 In.
All Models, Reverser Disconnect Clutch–All	Red	234 lbs. @ 1 11/16 In.

Check pressure plate for cracks, scoring or heat discoloration. Check cover and release levers for wear at pivot, distortion or other damage. On models with independent pto, check pto hub for wear or damage. PTO drive hub (damper) on diesel models is available

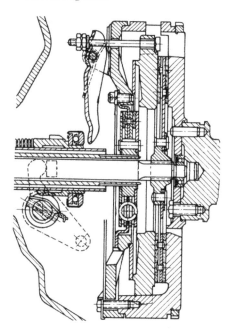

Fig. 111–Cross sectional view of single disc clutch shown exploded in Fig. 110.

Fig. 112–Measure release lever height from torsion damper hub as shown. Height should be 1.791-1.811 inches and equal to within 0.010.

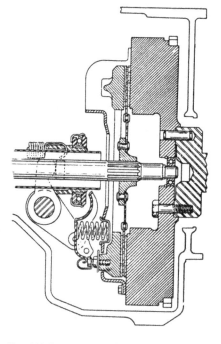

Fig. 114–Cross sectional, installed view of Auburn type single disc clutch assembly. Note that clutch spring pushes up against release lever instead of down against pressure plate when clutch is engaged.

separately. Release levers and pivot pins are available separately. Grind off peened ends of pivot pins for removal.

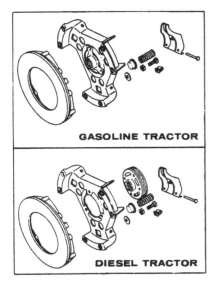

GASOLINE TRACTOR

DIESEL TRACTOR

Fig. 115–Exploded view of clutch cover used on some models with independent power take off. Remainder of assembly is similar to that shown in Fig. 113.

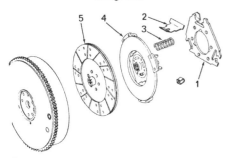

Fig. 116–Exploded view of Angle-Link type clutch used on Dubuque built Model 2030 with IPTO.

1. Spring ring
2. Angle clip
3. Spring
4. Cover assembly
5. Clutch disc

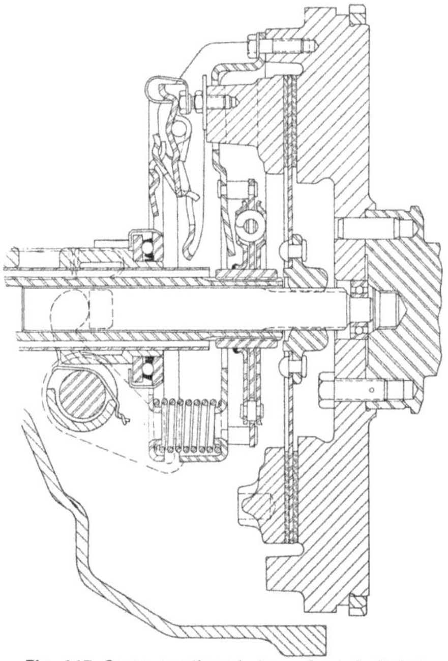

Fig. 117–Cross sectional view of angle-link type clutch shown exploded in Fig. 116.

Assemble the clutch by reversing the disassembly procedure. Adjust release levers to a height of 2.012 inches, measured between contact surface of lever and surface of flywheel. Finger height must not vary more than 0.010. John Deere Special Tool (JD-227) modified by grinding a small (0.156 inch deep) step in finger contacting surface, then using bottom of step to adjust finger height.

Angle-Link Type Clutch

114. Dubuque built Model 2030 tractors with independent power take-off are equipped with an angle-link type single disc clutch shown partially exploded in Fig. 116. A cross sectional view is shown in Fig. 117.

Clutch can be removed after clutch split as outlined in paragraph 111. When installing clutch, make sure long hub faces forward as shown in Fig. 117 and tighten clutch cover retaining cap screws to a torque of 35 ft.-lbs.

To disassemble the removed clutch cover, solidly block up under pressure plate on the bed of a press. Use a spacer to distribute the pressure evenly and compress spring ring (1—Fig. 116) until release levers are loose and angle clips (2) can be withdrawn. Remove all three angle clips and release pressure slowly, then lift off spring ring (1) and the six springs (3). If pressure plate and cover must be disassembled, mark the two parts with paint so balance can be maintained, remove the three adjusting screws and lift off the cover.

Springs (3) have a free length of 3 5/64 inches, and should test 118-130 lbs. when compressed to a height of 1 11/16 inches. On later models, torsion hub which drives the IPTO shaft is renewable. Release levers are a part of the cover and available only as an assembly.

Assemble by reversing the disassembly procedure. To adjust release levers after assembly, use a new clutch disc and install cover on flywheel as shown in Fig. 118. Using John Deere Clutch Adjusting Tool JD227, set all

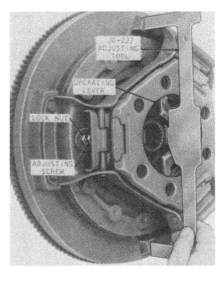

Fig. 118–A special adjusting tool (JD-227) is required to adjust release lever height.

three release levers to just touch adjusting tool center leg. Adjustment must be equal to within 0.010 inch. Reinstall return clips after adjustment is complete. Adjust clutch linkage as outlined in paragraph 109 after tractor is reconnected.

Series 1530, Dual Clutch

115. Refer to Fig. 119 for an exploded view of dual clutch unit and to Fig. 120 for cross sectional view. Clutch can be removed after clutch split outlined in paragraph 111. When installing clutch, make sure long hub of transmission clutch disc (2—Fig. 119) is forward. Use a suitable alignment tool and tighten retaining cap screws to a torque of 35-40 ft.-lbs.

Fig. 119–Exploded view of dual clutch unit used on Series 1530.

1. Flywheel
2. Transmission clutch disc
3. Front pressure plate
4. Diaphragm spring
5. Rear pressure plate
6. PTO clutch disc
7. Cover
8. Actuating bolt
9. PTO release pin
10. Release lever
11. Pivot pin
12. "E" ring
13. Adjusting nut
14. Jam nut

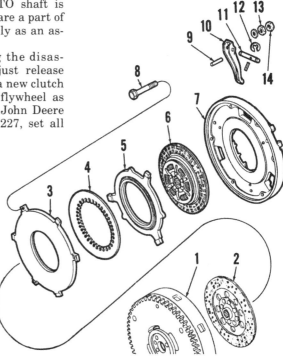

To disassemble the clutch cover and associated parts, remove locking nuts (14) and back off jam nuts (13) evenly until spring pressure is relieved. Mark the cover (7), rear pressure plate (5) and front pressure plate (3) with paint or other suitable means so balance can be maintained, and separate the units.

Inspect pressure plates (3 & 5) for cracks, scoring or heat discoloration and renew as necessary. Check diaphragm spring (4) for heat discoloration, distortion or other damage and renew if its condition is questionable. Renew transmission clutch disc (2) if facing wear approaches rivet heads, if hub is loose or splines are worn, or if disc is otherwise damaged. PTO disc (6) should be renewed if total thickness at facing area is 3/16-inch or less.

Assemble by reversing the disassembly procedure, installing pto clutch disc with hub offset toward front as shown in Figs. 119 and 120. Adjust release levers after installation as outlined in paragraph 110 and clutch linkage as in paragraph 109.

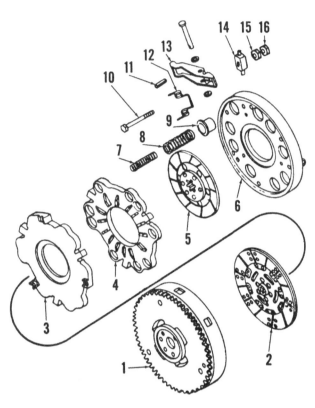

Fig. 121–Exploded view of dual clutch assembly used on most models so equipped. See Fig. 122 for cross section.

1. Flywheel
2. Transmission clutch disc
3. Front pressure plate
4. Rear pressure plate
5. PTO clutch disc
6. Clutch cover
7. Inner spring
8. Outer spring
9. Spring cup
10. Actuating bolt
11. PTO release pin
12. Spring
13. Release lever
14. Pivot block
15. Adjusting nut
16. Jam nut

Other Models, Dual Clutch

116. Refer to Fig. 121 for an exploded view and to Fig. 122 for cross section. Clutch can be removed after tractor split as outlined in paragraph 111. When installing clutch, make sure long hub of transmission clutch disc (2—Fig. 121) is forward. Use a suitable alignment tool and tighten retaining cap screws to a torque of 35 ft.-lbs.

To disassemble the removed clutch cover, place the assembly in a press and apply pressure to cover (6); or attach cover to a spare flywheel using equally spaced threaded rods and nuts.

Remove nuts (15 & 16) from operating bolts (10), then loosen the equally spaced nuts evenly until spring pressure is relieved.

Inspect pressure plates (3 & 4) for cracks, scoring or heat discoloration and renew as necessary. Clutch outer springs (8) are color coded red, should have an approximate free length of 3 1/16 inches, and should test 111 lbs. or more when compressed to a height of 1 7/8 inches. Inner springs (7) are color coded black. They should have an approximate free length of 3½ inches and should test 50 lbs. or more when compressed to a height of 1¾ inches. Renew springs which are distorted, heat discolored or fail to meet test specifications.

Assemble by reversing the disassembly procedure. Make sure long hub of clutch discs face away from each other as shown in Fig. 122. Tighten adjusting nuts (15—Fig. 121) evenly until release levers (13) just contact pto clutch push pins (11); back nuts off 2½ turns and secure by tightening jam nuts (16). Check lever adjustment after joining tractor, as outlined in paragraph 110, and equalize as necessary by making minor adjustments to high and low levers. Adjust clutch linkage as outlined in paragraph 109.

CLUTCH SHAFT

All Models

117. To remove clutch shaft it is necessary to separate clutch housing from transmission housing. Clutch shaft can be removed from rear of clutch housing.

If tractor has continuous-running pto and there is evidence of oil seepage between clutch shaft and powershaft (pto shaft) separate shafts and inspect oil seal and pilot. Press new pilot (cup rearward) into bore of powershaft until it bottoms. Press oil seal in bore of powershaft, with lip rearward, until it contacts pilot.

When installing clutch shaft, or

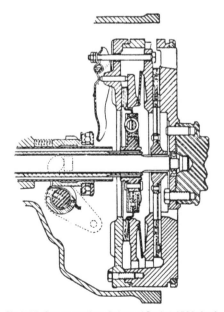

Fig. 120–Cross sectional view of Series 1530 dual clutch assembly showing clutch discs correctly installed.

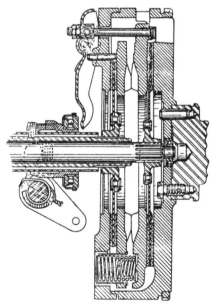

Fig. 122–Cross sectional view of dual clutch assembly. See Fig. 121 for exploded view and parts identification.

powershaft when tractor is equipped with continuous-running pto, be sure lugs on shaft align with slots in transmission oil pump drive gear.

CLUTCH RELEASE BEARING AND YOKE

All Models

118. The clutch release (throw-out) bearing (B—Fig. 123) can be removed after clutch housing is separated (split) from engine. Disconnect return spring and withdraw unit from carrier sleeve.

To remove yoke, disconnect clutch rod from clutch shaft arm, then drive out the two yoke retaining spring pins (RP). Pull clutch shaft out left side of clutch housing and catch yoke as it comes off clutch shaft.

When installing a new throw-out bearing, align index mark on bearing with notch in bearing carrier.

CLUTCH HOUSING

All Models

119. Clutch housing normally will not need complete removal for servicing. Clutch control linkage can be serviced when clutch housing is separated from engine. Clutch shaft and pto shafts along with their bearings and oil seals and the transmission oil pump can be serviced after clutch housing is

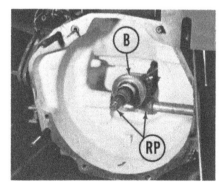

Fig. 123–View showing clutch throw-out bearing (B). Yoke is retained to clutch shaft by roll pins (RP).

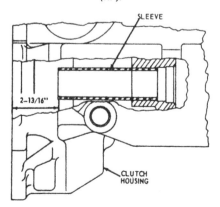

Fig. 124–Install clutch throw-out bearing carrier sleeve to dimension shown.

separated from transmission case.

Refer to paragraph 142 for service information on the transmission oil pump. Other service required on clutch housing will be obvious after examination and reference to the following: Clutch shaft bushing is installed with outer end flush with outer edge of bore. Clutch pedal pivot shaft is renewable and should be installed to protrude 2½ inches from housing. Clutch throwout bearing carrier sleeve is renewable and should be installed so that forward end is 2 13/16 inches from the machined engine mounting surface of clutch housing. (See Fig. 124). Oil seal for clutch shaft, or pto shaft, located in center of clutch housing can be renewed after removing shafts, transmission oil pump and needle bearing. Seal should be bottomed in bore with lip rearward. Needle bearing is also bottomed in its bore.

HI-LO SHIFT UNIT

Some tractors may be optionally equipped with a Hi-Lo Shift Unit which consists of a planetary gear set, hydraulically actuated direct drive clutch and hydraulically actuated multiple disc brake unit which locks the planet carrier to clutch housing. Pressure oil to operate the Hi-Lo Shift Unit is supplied by the transmission oil pump which also supplies lubrication for the transmission gears and excess flow supplies fluid for the main hydraulic system pump.

OPERATION

Models With Hi-Lo Shift

120. Refer to Fig. 125 for cross sectional view. Input shaft (1) is transmission clutch shaft and also carries hub for Hi-Clutch (2). When Hi-Clutch is engaged, planet carrier (3) is locked to input shaft providing a direct drive at crankshaft speed through Hi-Lo shift

Fig. 125–Schematic view of "Hi-Lo" Shift Unit showing main components.

1. Input shaft
2. Hi-Clutch
3. Planet carrier
4. Sun gear
5. Lo-Brake
6. Oil pump
7. PTO clutch shaft
8. PTO drive gears
9. Transmission input shaft

unit to transmission input shaft (9).

The hub for Lo-Brake unit (5) is splined to oil manifold hub. When Lo-Brake unit is engaged, planet carrier (3) is locked to clutch housing and power is transmitted through planet pinions to sun gear (4) and transmission input shaft at approximately ¾ crankshaft speed.

The spool-type control valve directs pressure oil to either the Hi-Clutch or Lo-Brake when dash-mounted control lever is moved. Refer to Fig. 126. The pressure regulating valve maintains 100-120 psi in control valve and the entire flow except for what leakage exists (through clutches or control valve) passes through the regulating valve to the main hydraulic pump.

TESTING

Models With Hi-Lo Shift

121. To check the system pressure, remove test plug (6—Fig. 127) from right side of shift cover and install a suitable pressure gage. Check the pressure with system at normal operating temperature and engine running at 2100 rpm. On 1020, 1520, 2020 and early 2030 Models, pressure should be 100-120 psi for models without Independent PTO; or 140-160 psi for models

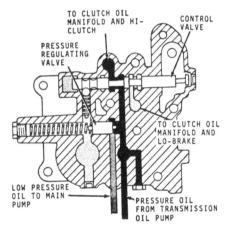

Fig. 126–Cross sectional view of control valve showing fluid passages.

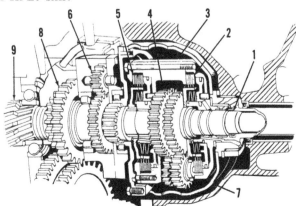

with Independent PTO. On all 1530 Models with late 2030 Models, pressure should be 125-135 psi with or without IPTO.

If regulated pressure cannot be properly adjusted, refer to paragraph 171 for additional hydraulic system checks and to paragraphs 122 and 123 for overhaul procedure.

OVERHAUL

Models With Hi-Lo Shift

122. CONTROL AND PRESSURE REGULATING VALVES. Control and pressure regulating valves are located in shifter cover as shown in Fig. 127. Regulating valve can be adjusted without removing shifter cover, but overhaul of valve unit can only be accomplished after cover is removed; proceed as follows:

Remove the four cap screws retaining transmission shield and remove the shield by working it up over shift lever boots. Disconnect Hi-Lo shift lever linkage and, if so equipped, independent pto lever link. Disconnect rear wiring harness. Remove clutch control lever cover (2) and, on models without independent pto, the cap screw hidden in control cover opening. Remove the remaining cap screws retaining shifter cover and lift off cover, valves and shift levers.

Pin (9) is inserted behind spring of valve detent (7). Remove plate (10) and

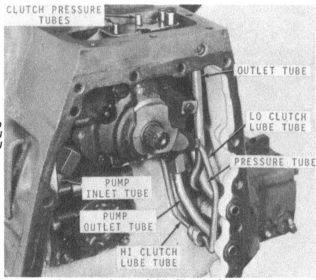

Fig. 128–Rear view of clutch housing showing clutch oil manifold, oil pump and oil lines.

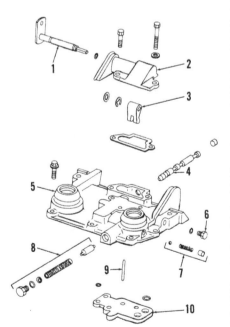

Fig. 127–Exploded view of shift cover and associated parts used on models with Hi-Lo Clutch unit. Refer to Fig. 176 for cover used on models equipped with IPTO option.

1. Valve shaft	6. Test plug
2. Housing	7. Detent assembly
3. Valve arm	8. Regulating valve
4. Valve spool	9. Detent pin
5. Shift cover	10. Cover

withdraw the pin before attempting to withdraw valve spool (4). Detent passage plug must be removed when reassembling, to depress the spring and reinstall pin.

Clean and inspect all parts. Control valve spool and regulating valve piston should slide freely in their bores without binding or excessive looseness. Detent valve spring should have an approximate free length of 15/16 inch and regulating valve spring a free length of approximately 3¼ inches.

123. HI-LO DRIVE UNIT. To remove the Hi-Lo drive unit, first split tractor between clutch housing and transmission as outlined in paragraph 136. Disconnect and remove hydraulic oil tubes from transmission oil pump and clutch housing (Fig. 128). Remove mid-pto if tractor is so equipped.

Unbolt and remove transmission oil pump and manifold assembly as a unit as shown in Fig. 129; then withdraw Hi-Lo Shift Unit as shown in Fig. 130.

To overhaul the removed unit, remove pto drive cover (26—Fig. 131) and withdraw planetary unit and associated parts. Remove the three long cap screws (through-bolts) from alternate holes in brake backing plate (25) and withdraw input shaft (13), Hi-Clutch drum (10) and associated parts from planetary unit. Remove the remaining three cap screws from backing plate (25) and withdraw Lo-Brake plates (23) and hub (24).

Thread a ¼-inch cap screw into planet pinion shaft (19) to act as a puller (Fig. 132) and withdraw the shaft, being careful not to lose locking balls (18—Fig. 131) at end of shaft. Lift out planet pinions (17) being careful not to lose the 44 loose needle bearings (16) as pinion is removed. Lift out sun gear (20) and thrust washer (21) after pinions are removed.

To remove pistons and springs (Belleville Washers) from planet car-

rier and Hi-Clutch drum, use a press and fixture as shown in Fig. 133. Collapse the spring washer pack and unseat snap ring, working through cutout portion of fixture. Refer to Fig. 134 for an exploded view. Install spring washers in pairs as shown in inset.

Hi-Clutch pack (11—Fig. 131) and Lo-Brake pack (23) both use the same separator plates and clutch discs, but one more disc is used in Lo-Brake pack. Externally splined steel separator plates in clutch or brake pack should be renewed if scored, warped, heat discolored or otherwise damaged. Stacked height of bronze-faced discs should be measured with a micrometer and renewed in sets, even though available individually. Renew the three bronze discs in Hi-Clutch pack if stack thick-

Fig. 129–Clutch oil manifold and transmission oil pump are removed as a unit.

THRUST WASHER

Fig. 130–Support "Hi-Lo" shift unit during removal. Note thrust washer located in bore of clutch housing.

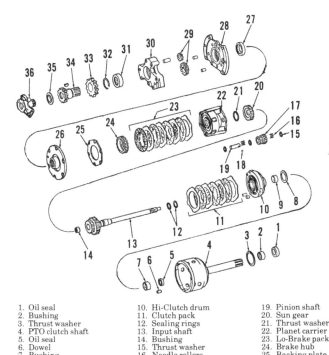

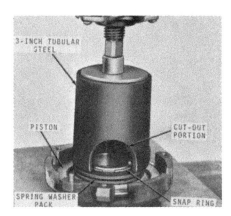

Fig. 133–Use a special tool as shown when compressing clutch spring pack to remove the retaining snap rings.

Fig. 131–Exploded view of Hi-Lo Clutch and drive unit showing component parts.

1. Oil seal	10. Hi-Clutch drum	19. Pinion shaft	28. Pump adapter
2. Bushing	11. Clutch pack	20. Sun gear	29. Pump gears
3. Thrust washer	12. Sealing rings	21. Thrust washer	30. Pump body
4. PTO clutch shaft	13. Input shaft	22. Planet carrier	31. Roller bearing
5. Oil seal	14. Bushing	23. Lo-Brake pack	32. Snap ring
6. Dowel	15. Thrust washer	24. Brake hub	33. PTO drive gear
7. Bushing	16. Needle rollers	25. Backing plate	34. PTO drive shaft
8. Thrust washer	17. Planet pinions	26. PTO drive cover	35. Thrust washer
9. Bushing	18. Locking ball	27. Roller bearing	36. Oil manifold

ness is less than 0.281. Renew the four bronze discs in Lo-Brake pack if stack thickness is less than 0.375.

Planet pinions (17) should be renewed in sets. When assembling planetary unit, install thrust washer (21) and output sun gear (20) in carrier with "V" marks on sun gear up. Assemble the two rows of 22 needle bearings in planet pinion (17) with spacer washer between bearing rows; then install planet pinion with "V" mark on face of pinion aligned with similar mark on sun gear. Push pinion shaft into housing until indentation in shaft is ready to enter carrier, install locking ball (18) and push shaft in until front end is flush with end of housing bore. Make sure "V" mark on all planet pinions properly align with mark on sun gear as pinions are installed.

Lo-Brake unit must be assembled and backing plate (25) installed before Hi-Clutch unit can be installed. Start with a steel separator plate next to piston and end with a bronze disc next to backing plate when assembling Lo-Brake unit. Start and end with a steel separator plate when assembling Hi-Clutch pack. Tighten cap screws and through-bolts to a torque of 23 ft.-lbs.

REVERSER UNIT

Some tractors may be optionally equipped with a Reverser Unit which consists of a planetary gear set, a hydraulically actuated forward drive clutch and hydraulically actuated reverse brake, all located in the clutch housing and operated by a dash mounted directional lever.

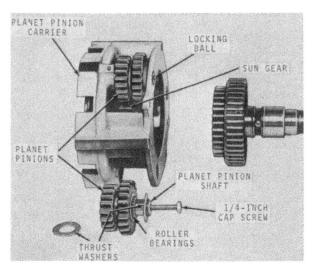

Fig. 132–Partially disassembled view of Hi-Lo Shift unit planet carrier and associated parts. Pinions must be timed when installed as outlined in paragraph 123.

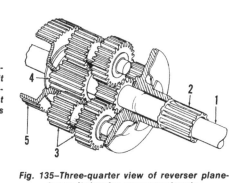

Fig. 134–Exploded view of Hi-Clutch drum and associated parts. Belleville spring washers are installed as shown in inset.

1. Thrust washer	5. Piston
2. Bushing	6. Piston ring
3. Hi-Clutch drum	7. Spring washers
4. O-ring	8. Snap ring

Pressure oil to operate the reverser unit is supplied by the transmission oil pump which also provides lubrication for the transmission gears and serves as a supply pump for the main hydraulic system pump.

OPERATION

Models With Reverser
124. Refer to Fig. 135 for a schematic view of planetary gear unit. Planet carrier drive shaft is splined into hub of forward clutch carrier. When forward

Fig. 135–Three-quarter view of reverser planetary unit showing component parts.

1. Clutch shaft	4. Secondary sun gear
2. Planet carrier	5. Reverse brake hub
3. Planet pinions	

clutch is engaged, clutch shaft is locked to carrier and the planetary unit turns together driving tractor in a forward direction.

The reverse brake unit is carried in transmission pump mounting plate and output sun gear (4) is splined into reverse brake hub. When reverse brake is applied, power entering at planet carrier causes the small planet pinion to walk around the fixed secondary sun gear (4) and the large planet pinion drives clutch shaft (1) at 16% overdrive ratio in a reverse direction.

The hydraulic control unit contains a shift valve which directs pressure fluid to either the forward clutch or reverse brake piston. No neutral position is provided in the shift valve. Also included is a clutch control valve which interrupts pressure and flow to the shift valve when clutch pedal is depressed, thus stopping forward or reverse motion of the tractor and permitting the gear change unit to be shifted in the normal manner. The design of the mechanical linkage prevents moving shift valve to reverse position when transmission range shifter lever is in "High" position.

Also included in the valve unit is a spring loaded accumulator piston which smooths the engagement during direction changes. An orifice screw is provided to permit adjustment of the shift rate. The accumulator remains charged when power flow is interrupted by the clutch pedal, thus permitting immediate and accurate control of clutch re-engagement rate.

CHECK AND ADJUST

Models With Reverser

125. Before making any tests, first check to make sure that transmission oil is at the correct level and that transmission is at operating temperature. It is generally advisable to check and adjust the unit in the order given.

126. **REGULATING VALVE.** Remove plug (29)—Fig. 136) and install a 0-300 psi pressure gage. With lefthand transmission shift lever in "Park" position and tractor running at high-idle speed, gage pressure should be 145-165 psi. To adjust the pressure, remove plug (1) and add or remove shim washers (7) as required.

127. **CLUTCH CONTROL VALVE.** First make sure regulating valve is correctly adjusted as outlined in paragraph 126, and record the pressure for later reference. Also check and/or adjust clutch pedal free play as outlined in paragraph 130.

Remove plug (13 or 14—Fig. 136) and install a 0-300 psi pressure gage. Move reverser control lever to pressurize the circuit containing the gage, then check circuit pressure which should equal system pressure as previously recorded.

To adjust the pressure, loosen clutch pedal clamp screw (2—Fig. 137). Turn pressure valve adjusting screw (3) until gage pressure reads 130 psi; turn screw in an additional 1½ turns and lock in place by tightening jam nut. Tighten clamp screw (2), making sure pedal is in upper position. If circuit pressure still does not equal system pressure, check for leakage in clutch or brake oil circuits.

128. **SHIFT ENGAGEMENT RATE.** Shift engagement rate when changing direction of travel can be adjusted within specified limits. Nominal shift time is ¾ to 1¼ seconds. Shift engagement rate can be varied by

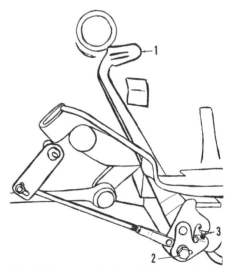

Fig. 137–Reverser clutch pedal linkage showing points of adjustment.

turning adjusting screw (30—Fig. 136) IN to slow shift rate or OUT to speed shift rate. Initial (average) setting is two (2) turns out from closed (bottomed) position. A slow shift rate accompanied by a jerky start could indicate a broken accumulator spring or sticking accumulator piston (26).

129. **HIGH SPEED LOCKOUT.** To adjust the high speed lockout, refer to Fig. 138. Move reverser control lever to "FORWARD" position and range shifter lever in "HIGH" position; then measure the clearance between lockout cam on reverser shift shaft and lockout pin as shown. Clearance should be 0.060; if it is not, disconnect link rod at lower end and turn adjusting yoke (Inset) as required until clearance is correct.

130. **PEDAL FREE PLAY.** When clutch pedal is pushed part way down, clutch control valve lowers reverser system pressure, stopping tractor motion. When pedal is fully depressed, flywheel clutch is released stopping

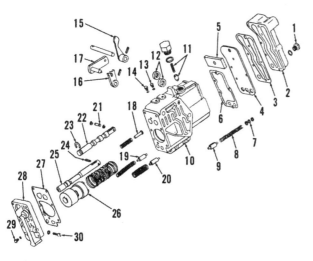

Fig. 136–Exploded view of reverser control valve showing component parts.

1. Plug	9. Regulating valve	16. Detent arm	23. Retaining ring
2. Cover	10. Body	17. Shift shaft	24. Pin
3. Gasket	11. Detent	18. Valve stop	25. Control valve
4. Plate	12. Oil seals	19. Cooler bypass valve	26. Accumulator piston
5. Block	13. Plug	20. Lubrication valve	27. Gasket
6. Gasket	14. Plug	21. Pin	28. Cover
7. Shims	15. Arm	22. Shift valve	29. Plug
8. Spring			30. Adjusting screw

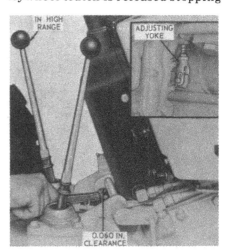

Fig. 138–Adjusting high speed lockout as outlined in paragraph 129.

power take-off and reverser drive shaft.

To adjust the free play, proceed as follows: With engine stopped, depress clutch pedal through first stage, until throwout bearing contacts clutch release levers, then measure distance from pedal pad to flywheel housing flange as shown in Fig. 139. Distance should be 6 inches on models before tractor serial number 80797, or 5¼ inches for later models. Adjust by disconnecting rod yoke and shortening or lengthening operating rod as required. Recheck clutch control valve as outlined in paragraph 127, after adjusting free play.

NOTE: Early clutch linkage can be identified by measuring length of slot in clutch fork (release) shaft. Slot is 1-inch long in early models and 1¼-inches in late units.

OVERHAUL

Models With Reverser

131. **REVERSER CONTROL VALVE.** To remove the reverser control valve, remove right platform and thoroughly clean brake valve, reverser control valve and surrounding area.

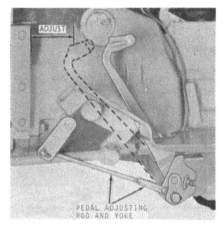

Fig. 139–Adjust clutch pedal free play to 6 inches on early models or 5¼ inches on late models by disconnecting yoke from pedal arm and changing length of rod.

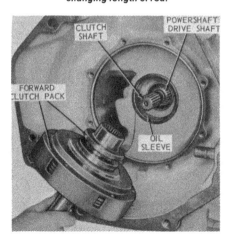

Fig. 140–Forward clutch pack can be removed as shown, after detaching engine from clutch housing.

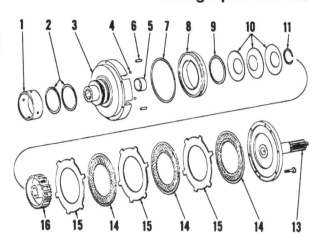

Fig. 141–Exploded view of forward clutch pack and associated parts.

1. Sleeve
2. Sealing rings
3. Clutch drum
4. Steel ball
5. Bushing
6. Dowel pin
7. Piston ring
8. Piston
9. Piston ring
10. Spring washers
11. Snap ring
13. Drive shaft
14. Clutch discs
15. Separator plates
16. Clutch hub

Disconnect brake lines and reverser link yoke. Drive out spring pin securing clutch shaft and control valve shaft. Remove the retaining cap screws and lift off brake valve and pedals. Remove the remaining three cap screws and lift off reverser control valve.

Refer to Fig. 136 for exploded view of reverser control valve. Evenly loosen cap screws retaining rear cover (28). Cover is under heavy pressure from accumulator spring when seated. Cooler bypass valve (19), lubrication valve (20) and pressure regulating valve (9) are interchangeable but springs are not. All parts are available individually and selective fitting is not required. Assemble by reversing the disassembly procedure. Tighten cover retaining cap screws to a torque of 70-75 inch pounds. Identify valve springs if necessary, by the following test data:

Regulating valve
(9) . . . . 30 lbs. @ 2 9/16 In.
Cooler bypass valve
(19) . . . . 20 lbs. @ 2 1/16 In.
Lubrication valve
(20) . . . . 4½ lbs. @ 13/16 In.
Clutch valve
(25) . . . . 32 Lbs. @ 1 ⅛ In.
Accumulator piston
(26) . . . . 310 lbs. @ 2 ⅝ In.

132. **FORWARD CLUTCH.** Forward clutch unit can be removed after splitting engine from clutch housing as outlined in paragraph 111. Remove clutch fork shaft and throwout bearing carrier and withdraw forward clutch pack as shown in Fig. 140.

Unbolt and remove clutch drive shaft (13—Fig. 141), then remove the three clutch discs (14) and separator plates (15). Lift out clutch hub (16). Refer to paragraph 134 for removal of clutch piston (8) and overhaul of forward clutch.

133. **REVERSE BRAKE AND PLANETARY UNIT.** To remove the reverse brake and planetary unit, first detach (split) clutch housing from transmission as outlined in paragraph

136. Transmission oil pump, reverse brake assembly and planetary unit can be removed as a unit as shown in Fig. 142; or oil pump may be removed separately by removing pump housing cap screws. To remove either unit, it is first necessary to remove mid-pto shift assembly and drive as outlined in paragraph 162.

Refer to Fig. 143 for an exploded view of reverse brake and planetary unit. Withdraw planet carrier (26), clutch shaft (23) and associated parts forward out of brake assembly. Withdraw clutch shaft to rear out of planet carrier.

Remove the three cap screws retaining carrier shaft (21) to carrier (26). Push each pinion shaft (35) forward slightly, remove and save locking balls (25), then remove shaft (35) and pinion (33), being careful not to lose the 44 loose needle rollers (32), thrust washers and spacer. Assemble by reversing the removal procedure. Note the stamped numbers on large gear teeth of pinions (33) and corresponding numbers on front of gear on clutch shaft (23). Time planetary unit during assembly by aligning corresponding numbers on clutch shaft and pinion.

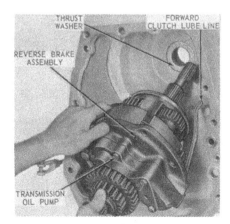

Fig. 142–Transmission oil pump, reverse brake and planetary unit can be removed as an assembly after detaching transmission from clutch housing.

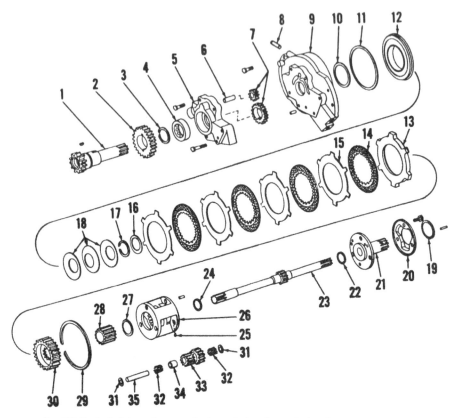

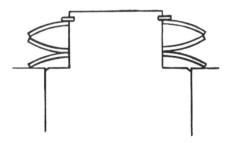

Fig. 146–Belleville spring washers on clutch and brake pistons must be positioned as shown.

Fig. 143–Exploded view of reverse brake and planetary unit showing component parts.

1. PTO drive shaft	9. Brake housing	18. Spring washers	27. Thrust washer
2. PTO drive gear	10. Sealing ring	19. Thrust washer	28. Secondary sun gear
3. Snap ring	11. Piston ring	20. Baffle	29. Snap ring
4. Bearing	12. Brake piston	21. Carrier shaft	30. Brake hub
5. Transmission pump	13. Backing plate	22. Thrust washer	31. Thrust washer
body	14. Brake discs	23. Clutch shaft	32. Needle rollers
6. Dowel pin	15. Separator plates	24. Thrust washer	33. Planet pinion
7. Pump gears	16. Thrust washer	25. Steel ball	34. Spacer
8. Lockout pin	17. Snap ring	26. Planet carrier	35. Planet pinion shaft

Square section sealing rings are used on clutch and brake units. Lubricate thoroughly and use care not to cut rings when installing. Position spring washers as shown in Fig. 146.

Four clutch discs (14—Fig. 141 or 143) are used in reverse brake; three in forward clutch. An equal number of separator plates (15) are used. Separator plates are slightly wavy and should not be flattened. Clutch (brake) pack begins with a separator plate next to actuating piston and ends with a clutch disc next to pressure plate. Clutch discs are 0.112-0.118 in thickness; separator plates 0.090. The three spring washers for forward clutch have slightly more cup than those for reverse brake.

TRANSMISSION

Overhaul reverse brake as outlined in paragraph 134.

Bushings in clutch housing are renewable. If new bushings are installed, make sure open ends of oil grooves are together as shown in Fig. 144. Tighten cap screws retaining carrier shaft (21—Fig. 143) to carrier (26) to a torque of 35 ft.-lbs. Tighten cap screws retaining reverse brake housing (9) to clutch

housing to a torque of 35 ft.-lbs. and cap screws retaining transmission oil pump to brake housing to 23 ft.-lbs.

134. **BRAKE/CLUTCH OVERHAUL.** A clutch piston disassembly tool can be made by cutting away a portion of a discarded oil filter cover as shown in Fig. 145; or from a suitable piece of pipe. Depress the spring washer pack in a vise as shown; unseat the snap ring then disassemble piston unit.

Transmissions are constant mesh type using helical cut gears. Two shift levers are used, the left lever selecting high, low and reverse ranges as well as a park position. The right lever controls a four step gear arrangement. Thus with the two shift levers, eight forward speeds (four in high range and four in low range) and four reverse speeds are available. The four reverse speeds approximate in mph the four forward speeds obtained in low range.

TOP (SHIFTER) COVER

All Models

135. **REMOVE AND REINSTALL.** To remove the shifter cover (except Low Profile), remove the shield and work it up over shifter lever boots. Disconnect connector from starting safety switch and bleed line from fitting on shift cover. If necessary, disconnect rear wiring harness, then unbolt and lift shifter cover from clutch housing.

Any further disassembly required will be obvious after examination of the unit. See Figs. 147 & 148.

Reinstall by reversing the removal procedure.

NOTE: Shifter rails and forks are an integral part of the transmission and

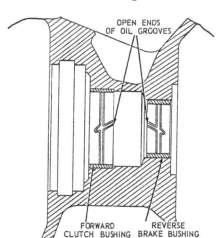

Fig. 144–Cross sectional view of clutch housing used on reverser models, showing bushings properly installed.

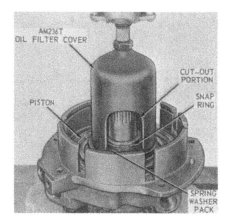

Fig. 145–An oil filter cover properly modified, makes a satisfactory disassembly tool for piston removal.

can be serviced after transmission is split from clutch housing as outlined in paragraph 136.

TRACTOR SPLIT

All Models

136. To split tractor between clutch housing and transmission case, disconnect battery ground straps, drain transmission case, remove shifter cover as outlined in paragraph 135, then remove the two clutch housing to transmission case cap screws located at rear of shifter cover opening under mounting flange. Remove left platform and unhook clutch return spring. Remove right platform, disconnect the two brake pressure lines from brake valve and the main hydraulic pressure line from tee or elbow near transmission oil filter. Remove plate retaining hydraulic pump inlet line and reservoir return line at lower right side of transmission case. Disconnect tail light wires and if so equipped, disconnect hydraulic outlet mid-couplers. Support transmission case, place a rolling floor jack under front section and block front axle to prevent front end from tipping. Remove remaining clutch housing to transmission cap screws and separate tractor. See Fig. 149.

NOTE: The cap screw (CS) located in front of the transmission filter (TF) cannot be completely removed unless filter is removed, however, cap screw can be unscrewed as tractor is split and left in casting hole. DO NOT lose the check valve, located in rear end of the main hydraulic inlet line, which will probably fall out as line comes out of transmission case.

When rejoining tractor, be sure spring is in place in forward end of pto shaft and ball is in place in rear end of mid-pto shaft, or mid-pto cover, if tractor is fitted with mid-pto or 540-1000 rpm pto. See Fig. 149.

SHIFTER SHAFTS AND FORKS

All Models

137. **REMOVE AND REINSTALL.** To remove shifter rails (shafts) and forks it will first be necessary to split tractor as outlined in paragraph 136. Remove seat and disconnect lift links from rockshaft arms. If so equipped, disconnect lines from rear of selective control valve and remove hose between selective control valve and rockshaft housing. Attach a hoist to rockshaft housing, place load selector lever in "L"

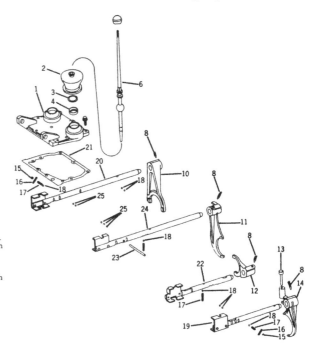

Fig. 147–Exploded view of shifter cover, shifter shafts and shifter forks. Shifter shafts are an integral part of transmission. See Fig. 149.

1. Shifter cover
2. Boot
3. Snap ring
4. Retainer
6. Shift lever
8. Set screw
10. Fork
11. Fork
12. Fork
13. Pin, starter safety switch
14. Fork
15. Plug
16. Spring pin
17. Detent spring
18. Detent ball
19. Shifter shaft, low range & rev.
20. Shifter shaft, 1st, 5th, 2nd, 6th
21. Gasket
22. Shifter shaft, high range
23. Interlock pin
24. Shifter shaft, 3rd, 7th, 4th, 8th
25. Interlock balls

position, then unbolt and remove rockshaft housing from transmission case.

Remove the starter safety switch pin from low range shifter fork, cut lock wires and remove set screws from shifter forks.

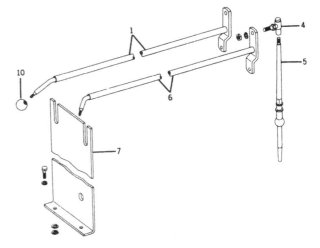

Fig. 148–Series 2020 Low Profile tractors have a gear shift lever arrangement as shown.

1. L.H. shifter lever
4. Ball joint
5. Gear shift lever
6. R.H. shifter lever
7. Shifter guide
10. Knob

Fig. 149–View showing front of transmission case after tractor split. Tractor shown is equipped with 540-1000 rpm pto with mid-pto.

CS. Cap screw
H. 1000 rpm gear
L. 540 rpm gear
S. Spring
SS. Shifter shafts
TF. Transmission filter

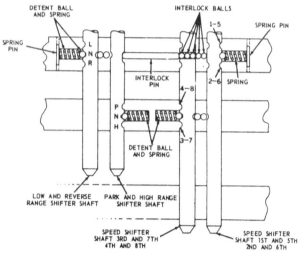

Fig. 150—Schematic view showing shifter shafts and the detent and interlock mechanisms. The rear detents are below shifter shafts instead of sides as indicated. Refer to text for service procedure.

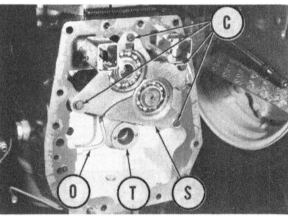

Fig. 151—Front of transmission case with pto shaft and pto gears removed. Countershaft support (S) must be removed before input shaft or pinion shaft can be removed.

NOTE: Before removal and installation of shifter shafts and forks, refer to Fig. 150 to determine location of detent and interlock mechanisms as well as for identification of shifter shafts. Also note that the five interlock balls are ¼-inch diameter whereas the remaining eight detent balls are 5/16-inch diameter.

Place the two left hand shifter shafts in neutral, then pull the outer range shift shaft (L N R) from its bore and use caution not to lose the three detent balls. Shift the inner speed shifter shaft into gear to release interlock pin,

then withdraw the inner range shifter shaft (P N H) from its bore and catch detent ball (and spring). Lift shifter

forks from transmission case.

Pull inner speed shifter shaft out of its bore until the three interlock balls can be removed from shaft, then complete removal of shaft and remove the three rear detent balls. Remove the remaining front two interlock balls from bore in housing. Pull the right hand speed shifter shaft from its bore and remove the remaining front detent ball. Lift shifter forks from transmission case.

NOTE: DO NOT turn shifter shafts while withdrawing from housing. Detent ball could drop into set screw hole, making shaft impossible to remove.

If interlock pin requires renewal, drive spring pin at right side of transmission case rearward and slide interlock pin out right side. NOTE: Detent bores are not aligned and interlock pin will not enter left detent bore.

Clean and inspect all parts and renew any that are bent, worn or otherwise damaged. Be sure to inspect all balls and renew any which have flat spots that would prevent them from rolling freely.

Reinstall shifter shafts and forks by reversing removal procedure. Start with right hand (outer) speed shifter shaft and shift inner speed shaft into a gear position before attempting to install right hand (inner) range shifter shaft so that interlock pin will move to the right. Secure all shifter shaft forks with lock wire.

1. Bearing cup	29. Shift collar
2. Bearing cone	30. Shift collar sleeve
3. Input shaft	31. 2nd & 6th gear
4. Needle bearing	32. Thrust washer (outer tangs)
5. Bearing cone	33. Retaining washer
6. Bearing cup	34. 4th & 8th gear
7. Shims	35. Thrust washer (thinnest)
8. Bearing quill	36. Shift collar
9. Shifter collar	37. Shift collar sleeve
10. Dowel	38. 3rd & 7th gear
11. Drive gear	39. Spacer
12. Dowel	40. Shims
13. Support	41. Bearing cup
14. Ball bearing	42. Bearing cone
15. Snap ring	43. Nut
16. Snap ring	44. Snap ring
17. Countershaft	45. Thrust washer
18. Brake spring	46. Reverse pinion
19. Brake plug	47. Shift collar
20. Ball bearing	48. Thrust washer
21. Snap ring	49. Shift collar sleeve
22. Snap ring	50. Low range pinion
23. Pinion shaft	51. Thrust washer
24. Bearing cone	52. Snap ring
25. Bearing cup	53. Snap ring
26. Shims	54. Needle bearing
27. 1st & 5th gear	
28. Thrust washer (thickest)	

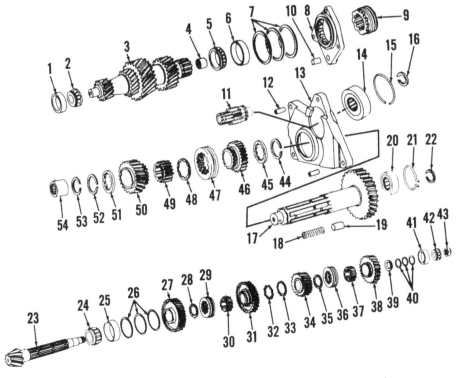

Fig. 152—Exploded view showing transmission shafts and gears. Note that transmission drive gear (11) is mounted in countershaft support (13).

COUNTERSHAFT

All Models

138. **R&R AND OVERHAUL.** To remove the countershaft, split tractor as outlined in paragraph 136 and remove shifter shafts and forks as outlined in paragraph 137. Remove the pto gear, or gears, from front of transmission. Remove cap screws from countershaft bearing support (13—Fig. 152). Remove snap ring (52) from its groove at rear end of countershaft, then use a screw driver and turn locking washer (48) until splines of washer index with splines of countershaft. Pry bearing support off dowels (12), pull assembly forward and lift gears from transmission case as they come off shaft. See Fig. 153.

Inspect all gears, thrust washers and shift collar for broken teeth, excessive wear or other damage and renew as necessary. If support assembly bearings or shafts, or the snubber brake assemblies (19—Fig. 152) require service, the shafts and bearings can be pressed out after removing retaining snap rings; however, the transmission drive gear must be removed before countershaft can be removed. Snubber brake springs (18) should test 63-77 lbs. when compressed to a length of 1½ inches. Needle bearing (54) can be removed from its bore after removing snap ring (53).

INPUT SHAFT

All Models

139. **R&R AND OVERHAUL.** To remove input shaft it is first necessary to remove countershaft as outlined in paragraph 138.

With countershaft removed, the transmission input shaft is removed as follows: Remove transmission oil cup and lines. Straighten lock plates and

Fig. 154—View of transmission oil pump (P) with clutch shaft and pto powershaft removed.

I. Inlet line
O. Outlet line
P. Transmission pump

remove input shaft bearing quill (8) and shims (7) from front of input shaft. Bump input shaft forward until front bearing cup (6) clears its bore, then move input shaft forward, lift rear end of shaft and remove input shaft from transmission case.

With input shaft removed, inspect all gears for chipped teeth or excessive wear. Inspect bearings and renew as necessary. Bump bearing cup (1) forward if removal is required. Inspect needle bearing (4) and renew if necessary.

Install and adjust end play of input shaft as follows: Be sure bearing cup (1) is bottomed in bore and place input shaft in position. Use original shim pack (7), or use a new shim pack approximately 0.030 thick, install front bearing quill (8) and tighten cap screws to 35 ft.-lbs. torque. Use a dial indicator to check the input shaft end play which should be 0.004-0.006. Vary shims as required. Shims are available in thicknesses of 0.003, 0.005 and 0.010. Do not forget front oil line clamp when making final installation.

PINION SHAFT

All Models

140. **R&R AND OVERHAUL.** To remove the transmission pinion shaft, remove the input shaft as in paragraph 139 and the differential as in paragraph 144.

With differential removed, remove oil line, nut (43—Fig. 152), bearing (42), shims (40) and spacer (39). Use a screw driver and turn thrust washers until splines of thrust washers are indexed with splines of countershaft. Pull countershaft rearward and remove parts from transmission case as they come off shaft. Bearing cup (25) and shims (26) can be removed from

housing by bumping cup rearward. Be sure to keep shims (26) together as they control the bevel gear mesh position. Bearing cup (41) can be removed from housing by bumping cup forward.

Check all gears and shafts for chipped teeth, damaged splines, excessive wear or other damage and renew as necessary. If pinion shaft is renewed, it will also be necessary to renew the differential ring gear and right hand differential housing as these parts are not available separately. Bearing (24) is installed with large diameter toward gear end of shaft.

NOTE: Mesh (cone point) position of the pinion shaft and main drive bevel pinion gear is adjusted with shims (26) located between rear bearing cup (25) and housing. If new drive gears or bearings are installed, the mesh position must first be checked and adjusted as outlined in paragraph 147. If same pinion shaft and bearing are installed, reinstall the same shims (26) and check the bearing preload as in paragraph 141.

Install pinion shaft and adjust shaft bearing preload as follows: Use Fig. 152 as a guide and with bearing (24) on pinion shaft, start shaft into rear of housing. With shaft about half-way into housing, place 1st and 5th speed gear (27) on shaft with teeth for shift collar (29) toward front. Place the thickest thrust washer (28) on shaft, then install coupling sleeve (30) and shift collar (29). Move shaft forward slightly and install 2nd and 6th speed gear (31) with teeth for shift collar toward rear. Place thrust washer with outer tangs (32) over shaft, then slide retaining washer (33) over thrust washer (32). Move shaft slightly forward and install 4th and 8th speed gear (34) on shaft with teeth for shift

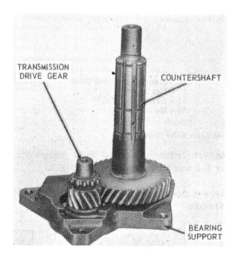

TRANSMISSION DRIVE GEAR

COUNTERSHAFT

BEARING SUPPORT

Fig. 153—View of removed countershaft bearing support assembly.

collar toward front. Place the thinnest thrust washer (35) on shaft and install shift collar sleeve (37) and shift collar (36). Install 3rd and 7th speed gear (38) on shaft with teeth for shift collar toward rear. Push shaft forward until rear bearing cone (24) seats in bearing cup (25) and use screw driver to turn thrust washers until splines on thrust washers lock with splines of pinion shaft. Install spacer (39), shims (40), bearing (42) and nut (43), then adjust pinion shaft bearing preload as outlined in paragraph 141.

141. PINION SHAFT BEARING ADJUSTMENT. The pinion shaft bearings must be adjusted to provide a bearing preload of 0.006 (5-15 in. lbs. rolling torque). Adjustment is made by varying the number of shims (40—Fig. 139).

To adjust the pinion shaft bearing preload, proceed as follows: Mount a dial indicator with contact button on front end of pinion shaft and check for end play of shaft. If shaft has no end play, add shims (40) to introduce not more than 0.002 shaft end play. NOTE: Do not exceed more than 0.002 shaft end play when beginning adjustment as increased end play increases the possibility of inaccuracies due to parts shifting. If original shims (40) are not being used, install a preliminary 0.035 thick shim pack. Shims are available in thicknesses of 0.002, 0.005 and 0.010. Tighten nut (43) to 160 ft.-lbs. torque and measure shaft end play, then remove shims from shim pack (40) equal to the measured shaft end play, PLUS an additional 0.005. This will give the recommended bearing preload of 0.006. Retighten nut to 160 ft.-lbs. torque and stake in position.

TRANSMISSION OIL PUMP
All Models

The Transmission oil pump is a gear type pump mounted on rear wall of clutch housing and is driven by the engine clutch shaft on tractors without pto, or with transmission-driven pto. On tractors with continuous-running pto, pump is driven by the pto powershaft. See Fig.154.

Pump capacity is 6 gpm @ 2500 rpm. The main hydraulic pump and power steering system have priority on the output of transmission oil pump. Models equipped with open-center steering systems are equipped with a separate pump. Most of the main hydraulic pump fluid is routed to the trasmission as hydraulic charge oil. Depending on the systems installed onto the year/model tractor (such as hi/lo shift or independent pto), hydraulic oil is routed from the pump using separate passages and hydraulic lines. When reservoir is filled, the overflow on some models tractors returns to lubricate the transmission and to fill the brake valve reservoir. On tractors with oil cooler, the overflow is passed through an oil cooler mounted on right side of radiator before it returns to the transmission and brake control valve. Oil cooler is shown in Fig. 82.

Transmission oil pump system is protected by a 100 psi relief valve located in clutch housing directly back of brake control valve. See Fig. 155.

142. TESTING. The operation of the transmission oil pump can be tested as follows: Remove side screens and hood, then disconnect hydraulic reservoir outlet hose from return line or oil cooler. Attach a line fitted with a shut-off valve and a pressure gage of at least 300 psi capacity to the detached end of reservoir outlet hose. Position a clean container to catch flow of oil from shut-off valve. Place transmission in "PARK" position, depress clutch pedal, then start engine and run at 2500 rpm. Release clutch and check flow of pump which should be one gallon in ten seconds. Now close shut-off valve until gage pressure raises to 50 psi at which time the pump should still produce one gallon every ten seconds.

If pump will meet above conditions it can be considered satisfactory. If pump will not meet above conditions, clean the transmission oil filter and the inlet screen, which is located under the square headed plug just behind transmission oil filter, then retest pump. If pump operation is still not satisfactory, check to see that the relief valve, lo-

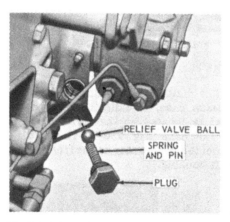

Fig. 155–Relief valve assembly for transmission oil pump is located as shown.

cated in clutch housing directly back of brake control valve (Fig. 155), is not stuck in the open position. Also check relief valve spring which should test 30½-37½ lbs. when compressed to a length of 1½ inches. If relief valve seat is damaged, seat can be pried out and a new one driven in.

If performing the above operations does not produce proper pump operation, remove and overhaul pump as outlined in paragraph 143.

143. R&R AND OVERHAUL. To remove the transmission oil pump, first split clutch housing from transmission housing as outlined in paragraph 136.

With clutch housing separated from transmission case, pull clutch shaft, or clutch shaft and pto power shaft, from clutch housing. Remove pump inlet and outlet lines from pump, then remove pump from wall of clutch housing and separate pump body from adapter. See Fig. 156.

Clean and inspect all parts for chipping, scoring or excessive wear. If bearing in pump body requires renewal, press new bearing in bore until it bottoms. Pump gears are available as a matched set only. Pump idler shaft is renewable and diameter of new shaft is 0.6240-0.6250. Thickness of new pump gears is 0.5085-0.5095.

When reassembling pump, coat gears with oil and tighten adapter mounting cap screws to 35 ft.-lbs. torque. Align slots of clutch shaft, or pto powershaft, with lugs of pump drive gear when installing shafts and be sure seals are on ends of inlet and outlet tubes before rejoining tractor.

DIFFERENTIAL
AND
FINAL DRIVE

Differential units in series 1020 tractors have two pinions while other models have four. Both differentials may be equipped with a differential lock. Removal procedure will be the same for all tractors.

Final drives incorporate a planetary gear reduction unit at inner end of the

Fig. 156–View showing transmission oil pump separated. Note drive lugs in I. D. of pump drive gear.

housing and final drive units for all models are similar. See Fig. 159.

DIFFERENTIAL

All Models

144. **REMOVE AND REINSTALL.** To remove differential, drain transmission case, then remove final drives as outlined in paragraph 150 and the rockshaft housing at outlined in paragraph 182.

If tractor is equipped with a differential lock, the assembly must be removed as follows: Remove clamp screw from lever (1—Fig. 157), or pedal (12) on some models, and remove the square key (9). Hold the yoke (4) in place, bump shaft (3) rearward and remove woodruff key (8) when it clears yoke. Continue to bump shaft rearward until plug at rear end of shaft bore comes out, then remove shaft, yoke and collar (7).

Disconnect load control arm spring, slide pivot shaft to the left and lift out load control arm. Remove transmission oil cup and rear oil line. Support the differential, remove both bearing quills (2 & 11—Fig. 158) and keep shims (3) with the correct quill.

The shims (3) located between bearing quills and transmission case control the differential bearing preload and the backlash of main drive bevel gears. Recommended bearing preload is 0.002-0.005 and recommended backlash is 0.006-0.012. Refer to paragraph 146 for adjustment procedure.

145. **OVERHAUL.** On all models the bevel ring gear, right hand differential housing and pinion shaft are available as a matched set.

To disassemble unit, remove the eight differential housing bolts and separate housing. Note that the single bevel pinion shaft of 1020 differential is located by a dowel pin.

Removal of bearings (5 and 13—Fig. 158), and bearing cups (4 and 14) is obvious. If any of the pinions (9) or pinion shafts (10) are damaged, all mating parts should also be renewed. If

axle (side) gears (8) are damaged or excessively worn, closely examine bores in differential housing as they may also be damaged.

Reassemble by reversing the disassembly procedure and tighten differential housing cap screws to 35 ft.-lbs. Cap screws are self locking.

MAIN DRIVE BEVEL GEARS

All Models

146. **ADJUSTMENT.** If differential is removed for access to other parts and no defects in the adjustments are noted, the shim packs should be kept intact and reinstalled in their original positions. However, if bevel gears, bearings, bearing quills or transmission case are renewed, the main bevel gears should be checked for mesh (cone point) position, differential carrier bearing preload and gear backlash, and in the foregoing order.

147. **MESH (CONE POINT) POSITION.** The fore and aft position of the transmission pinion shaft is controlled by shims located between pinion shaft rear bearing cup and front wall of differential compartment as shown in Fig. 160. When renewing parts the shim

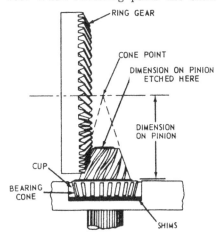

Fig. 158–Exploded view of series 2020 differential. Other models similar except only one pinion shaft (10) is used on 1020.

2. Quill, LH	5. Bearing cone	9. Pinion	12. Bevel gear set
3. Shims	7. Housing, LH	10. Pinion shaft	13. Bearing cone
4. Bearing cup	8. Bevel (side) gear	11. Quill, RH	14. Bearing cup

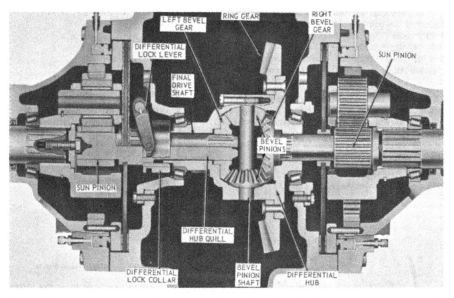

Fig. 160–Simplified view of main drive bevel gears showing location of shims which control gear mesh position. Refer to text for adjustment procedure.

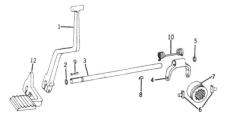

Fig. 157–Exploded view showing parts of differential lock. Some models use pedal (12) instead of lever (1).

1. Lever	
2. "O" ring	7. Collar
3. Shaft	8. Woodruff key
4. Yoke	9. Square key
5. Plug	10. Spring
6. Shoe	12. Pedal

Fig. 159–Cross-sectional view showing the differential, differential lock and planetary reduction units of a series 1020 tractor. Other units are basically similar.

pack required for correct mesh position of the transmission pinion shaft and the main drive bevel gear can be determined as follows: Use 1.446 inches as the nominal width of an assembled rear pinion shaft bearing and to the 1.446 inches, add the dimension found etched on aft end of pinion shaft gear. Next, observe the actual cone point dimension stamped on top rear of transmission case, then subtract the previously determined value from the cone point dimension. This will give the required thickness of shim pack. The bearing assembly width (1.446), plus dimension on end of pinion shaft, plus shim pack, should equal the dimension stamped on top rear of transmission case.

As an example, assume the following values: 6.643 stamped on rear of transmission case and 5.183 etched on aft end of pinion shaft gear. Compute shim pack thickness as follows: add 5.183 and 1.446 which equals 6.629, then subtract 6.629 from 6.643 which leaves 0.014 which will be the shim pack thickness required for correct mesh position of the main drive bevel gears. Follow this procedure whenever any new parts are installed that affect mesh position of the bevel gears.

148. DIFFERENTIAL BEARING ADJUSTMENT. The differential carrier bearings should have a preload of 0.002-0.005 and adjustment is made as follows: Install differential and bearing quills with original shim packs, then check differential end play using a dial indicator.

NOTE: When making this adjustment, be positive that clearance exists between the main drive bevel ring gear and pinion shaft at all times.

If no differential end play exists, add shims under right bearing quill to introduce not more than 0.002 end play. If more than 0.002 end play existed on original check, subtract shims.

Measure end play of differential, then subtract shims equal to the measured end play PLUS an additional 0.003 to give the desired 0.002-0.005 bearing preload. Shims are available in thicknesses of 0.003, 0.005 and 0.010.

149. BACKLASH ADJUSTMENT. With differential carrier bearing preload adjusted as outlined in paragraph 148, adjust backlash between main drive bevel gear and pinion shaft to 0.006-0.012 by transferring bearing quill shims from one side to the other as required. Moving shims from left to right will decrease backlash. Do not remove any shims during backlash adjustment or the previously determined preload adjustment will be changed.

FINAL DRIVE

All Models

150. REMOVE AND REINSTALL. To remove final drive, support rear of tractor and remove wheel and tire. Disconnect fender lights and free wiring harness from clamp on final drive housing, then remove fender. If right hand final drive is being removed and tractor has selective control valve, disconnect pressure line, coupler lines and return hose between valve and rockshaft housing and remove control valve. Disconnect the brake line from final drive housing. Attach hoist to final drive, remove attaching cap screws and pull final drive from transmission case.

Reinstall by reversing removal procedure and tighten attaching cap screws to 85 ft.-lbs. torque.

NOTE: If brake disc came off with final drive, install disc so that thickest facing is toward transmission case.

151. OVERHAUL. To overhaul the removed final drive unit, refer to Fig. 161 and proceed as follows:

Remove lock plate (23), cap screw (24) and retainer washer (25), then pull planet carrier assembly (26) from axle. Support outer end of final drive housing (11) so oil seal (2) will clear and press axle out of housing. The axle bearings, bearing cups and oil seals are now available for inspection or renewal. If outer bearing (4) is renewed, heat it to approximately 300°F. and drive it into place while hot. NOTE: If axle is flanged, be sure oil seal is on axle, metal side out, before installing outer bearing. Bearing cups are pressed in bores until they bottom. Seal

cup (3) will be pushed out when outer bearing cup is removed. Be sure to reinstall seal cup after bearing cup is installed. If ring gear and/or final drive housing is damaged, renew complete unit.

To remove planet pinions (20), expand snap ring (27), lift it from groove of carrier (26) and pull pinion shafts (18). Check carrier, pinions and rollers for pitting, scoring or excessive wear and renew parts as required. If any of the planet pinion rollers are defective, renew the complete set.

Reassemble final drive and adjust axle bearings as follows: Coat bores of planet pinions with grease and position rollers (23 in each bore) in pinions. Place a thrust washer on each side of pinions, then place pinions in carrier and insert pinion shafts only far enough to retain rollers and thrust washers. Install snap ring (27) in slots of pinion shafts, then complete insertion of pinion shafts and be sure snap ring seats in groove in carrier. Coat inner seal (13) with grease and install axle in housing. Heat inner bearing (15) to approximately 300°F. and install bearing on inner end of axle. Place carrier assembly on axle, install retaining washer (25) and cap screw (24) and tighten cap screw until bearing is pulled into place and a small amount of axle end play remains. Now while bearing is still hot, check the amount of torque required to turn the axle with the existing axle end play, then tighten the cap screw to increase the rolling torque 50-80 in.-lbs. for old bearings, or 90-140 in.-lbs. if new bearings are used. Install lock plate (23). Fill axle outer bearing opening with multi-pur-

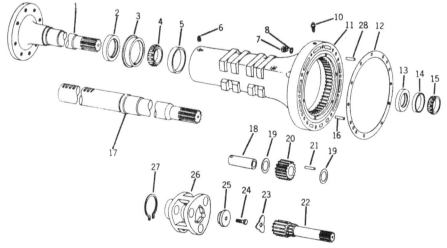

Fig. 161—Exploded view of final drive assembly. All units are similar although all parts are not interchangeable.

1. Axle, flanged	8. "O" ring	15. Bearing cone	22. Final drive shaft
2. Oil seal	10. Bleed screw	16. Dowel pin	23. Lock plate
3. Oil seal cup	11. Housing	17. Axle, adj.	24. Cap screw
4. Bearing cone	12. Gasket	18. Pinion shaft	25. Retaining washer
5. Bearing cup	13. Oil seal	19. Thrust washer	26. Carrier
6. Plug	14. Bearing cup	20. Pinion	27. Snap ring
7. Plug		21. Rollers (69 used)	28. Dowel (replacement)

pose grease and install oil seal with metal side out.

Use new gasket (12) when reinstalling final drive to transmission case. However, before installing final drive, pull final drive shaft (22) and brake disc and inspect. Brake disc is installed with thickest facing next to transmission case.

Refer to paragraph 159 for information on brake pressure plate and pressure ring.

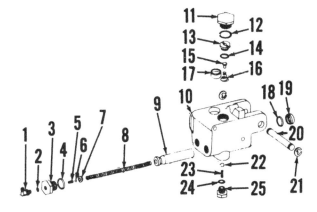

Fig. 163—Exploded view of brake control valve. Only half of component parts are shown. Other half of parts are identical.

BRAKES

The brakes on all models are hydraulically actuated and utilize a wet type disc controlled by a brake operating valve located on right side of clutch housing. See Fig. 162. Brake discs are splined to the final drive shafts and the brake pressure ring is fitted in inner end of final drive housing. Except for a pedal adjustment, no other brake adjustments are required.

1. Connector	8. Spring
2. "O" ring	9. Piston
3. Check valve seat	10. Housing
4. "O" ring	11. Filler plug
5. Spring	12. "O" ring
6. Ball	13. Valve seat
7. Retainer	
14. "O" ring	20. Pedal shaft
15. Check valve	21. Retainer ring
16. Spring	22. Ball
17. Cup plug	23. Spring
18. "O" ring	24. "O" ring
19. Oil seal	25. Plug

BLEED AND ADJUST

All Models

152. **BLEEDING.** Brakes must be bled when pedals feel spongy, pedals bottom, or after disconnecting or disassembling any portion of the braking system.

To bleed brakes, start engine and run for at least two minutes at 2000 rpm to insure that brake control valve reservoir is filled.

NOTE: Brakes can also be bled without engine running if necessary. Remove right platform and fill brake control valve reservoir by removing filler plug (11—Fig. 163). Follow same bleeding procedure used when engine is running except brake valve reservoir will need to be refilled after each fifteen strokes of the brake pedal.

Attach a bleeder hose (preferably clear plastic) to brake bleed screw located on top side of final drive housing and place opposite end in filler hole of rockshaft housing. Slowly depress and release brake pedal until oil flowing

from bleeder hose is completely free of air bubbles, then depress brake pedal and tighten bleed screw.

Repeat bleeding operation for opposite side brake.

153. **ADJUSTMENT.** Whenever brake control valve has been disassembled, a brake pedal and equalizing valve adjustment must be made to prevent mechanical interference between brake valve pistons and reservoir check valves.

Before making this adjustment, bleed brakes as outlined as paragraph 152.

154. **RIGHT PEDAL.** Adjust right hand pedal stop screw so brake valve piston is fully extended and arm of brake pedal is snug against end of piston without piston being depressed (zero clearance). Apply a force of about 10 lbs. to LEFT brake pedal and if left brake pedal settles, turn pedal stop screw for right brake pedal counter-clockwise about ⅓-turn at which time left brake pedal should stop settling. If left brake pedal does not stop settling, a leak in the braking system is indicated and must be isolated and corrected. Refer to paragraph 157.

155. **LEFT PEDAL.** Adjust left hand pedal stop screw so brake valve piston

is fully extended and arm of brake pedal is snug against end of piston without piston being depressed (zero clearance). Apply a force of about 10 lbs. to RIGHT brake pedal and if right pedal settles, turn pedal stop screw for left brake pedal counter-clockwise about ⅓-turn at which time right brake pedal should stop settling. If right brake pedal does not stop settling, a leak in the braking system is indicated and must be isolated and corrected. Refer to paragraph 157.

156. **PEDAL HEIGHT.** If brake pedal height is not aligned after equalization valves are adjusted as outlined in paragraphs 154 and 155, align pedals by turning stop screw on highest pedal about 1/6-turn counter-clockwise.

BRAKE TEST

All Models

157. **PEDAL LEAK-DOWN.** With a 60 lb. pressure applied continuously to each pedal for one minute, the pedal leak-down should not exceed one inch. Excessive brake pedal leak-down can be caused by air in the brake system, faulty brake control valve pistons and/or "O" rings, faulty brake pressure ring seals, or faulty brake control valve equalizing valves or reservoir check valves.

Brakes should always be bled as outlined in paragraph 152 before any checking or adjusting of braking system is attempted. Faulty brake control valve pistons or "O" rings will be indicated by external leakage around the brake control valve pistons.

Faulty brake pressure ring seals, or brake control valve, can be determined as follows: Isolate brake from brake control valve by plugging brake line. If leak-down stops, the brake pressure ring seals are defective. If leak-down

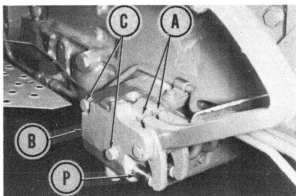

Fig. 162—View showing brake control valve. Right platform has been removed.

A. Pedal adjusting screws
B. Brake control valve
C. Mounting cap screws
P. Brake valve pistons

continues, the brake control valve is faulty and can be checked further by depressing brake pedals individually, then simultaneously. If leak-down occurs in both cases, a defective reservoir check valve is indicated. If leak-down occurs during individual pedal operation but not on simultaneous pedal operation, a faulty equalizer valve is indicated.

Refer to paragraph 158 for brake control valve information and to paragraph 159 for brake pressure ring information.

OVERHAUL

All Models

158. **BRAKE CONTROL VALVE.** To remove brake control valve, remove right platform and thoroughly clean valve and surrounding area. Disconnect brake lines from rear of control valve, remove the mounting cap screws (C—Fig. 162) and remove control valve from clutch housing. Discard gasket located between control valve and clutch housing. Remove "E" ring (21—Fig. 163), pull shaft (20) and remove pedals from control valve. Remove connectors (1), check valve springs (5) and balls (6). Remove seats (3) and ball retainers (7), then push pistons (9) and springs (8) out rear of valve body. Remove filler plug (11) and cup plug (17), then using a screw driver with proper sized bit, remove reservoir check valve assemblies (items 13, 14, 15 & 16). Remove equalizer valve assemblies (items 22, 23, 24 & 25). "O" rings (18) and oil seals (19) can be removed from piston bores.

Clean and inspect all parts. Piston spring (8) should test 20 lbs. when compressed to a length of 5¾ inches. Renew housing (10) if seats for equalizer balls (22) are damaged. Oil seals (19) are installed with lips toward outside. Pay particular attention to area of reservoir check valve (15) where contact is made with valve piston and

renew valve if any doubt exists as to its condition. Brake pedals are fitted with bushings for brake pedal shaft (20) and bushings and/or shaft should be renewed if clearance is excessive.

Lubricate lips of oil seals (19) and all other parts. Use a new cup plug (17) and reassemble by reversing disassembly procedure. Use a new gasket when installing valve on tractor. Bleed brakes as outlined in paragraph 152 and adjust pedals as outlined in paragraphs 154, 155 and 156.

159. **BRAKE PRESSURE PLATE, RING AND DISC.** To remove brake pressure plate, pressure ring and brake disc, remove final drive housing as outlined in paragraph 150. Pull final drive shaft from differential and remove brake disc from final drive shaft. See Fig. 164. Lift brake pressure plate from dowels in final drive housing. Remove brake pressure ring by prying it out evenly. If pressure ring is difficult to remove, attach a small hydraulic pump to brake line connections, be sure bleed valve is closed, then pump oil behind pressure ring to force it from cylinder (groove). See Fig. 165. Dowels can be removed from final drive housing, if necessary.

Inspect brake disc for worn or damaged facing or damaged splines. If facings require renewal, renew complete disc assembly as facings are not available separately. Inspect pressure plate for scoring, checking, or other damage and renew if necessary. Remove and discard seals from pressure ring and inspect ring for cracks or other damage.

To reassemble brake assembly, proceed as follows: Place brake disc on final drive shaft so thickest facing is next to transmission case and insert final drive shaft into differential. Place

new inner and outer seals on pressure ring and lubricate assembly liberally. Start pressure ring into its cylinder (groove) with flat chamfered side first and press into cylinder until it bottoms. Be absolutely sure that neither seal is cut or rolled during installation. Place pressure plate over dowels in final drive housing, hold in place if necessary, then install final drive housing.

Bleed brakes as outlined in paragraph 152.

POWER TAKE-OFF (Continuous Or Transmission)

Tractors may be equipped with 540 rpm transmission driven pto, a 540 rpm continuous running pto or a dual 540-1000 rpm continuous running pto. In addition, a mid-pto is also available with either of the continuous running pto units, however, the mid-pto operates only at 1000 rpm.

Changing speeds of the dual speed pto is accomplished by using either a six spline 540 rpm stub (output) shaft or a 21 spline 1000 rpm stub (output) shaft. The 540 rpm stub shaft incorporates a pilot on its forward end which moves the pto shaft assembly forward against spring pressure to engage splines on pto shaft with the splines of the 540 rpm pto driven gear. When the 1000 rpm pto stub shaft, with no pilot, is installed, the spring in forward end of pto shaft assembly moves the pto shaft rearward and the splines of the pto shaft engage with the splines of the 1000 rpm pto driven gear. See Fig. 166.

Shifter couplings are provided to engage and disengage rear pto and mid-pto shafts.

When service is required on the pto system, the following should be taken into consideration. Work involving rear pto shaft can be done by working from rear of tractor. Work involving the pto driven gears, powershaft clutch shaft, mid-pto shaft or mid-pto shifter assembly will require that the clutch housing be separated from the transmission case. Work involving the rear pto shaft shifter assembly will involve removing the rockshaft housing, high and low range shifter shafts and countershaft assembly in addition to separating the clutch housing from transmission case.

Fig. 164–Final drive shaft and brake disc being removed. Thickest disc facing is next to transmission case.

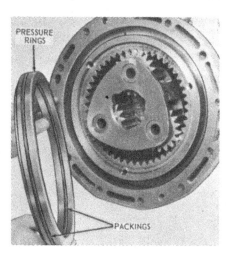

Fig. 165–View showing pressure ring removed from cylinder (groove) in final drive housing. Note that flat chamfered side is toward bottom of cylinder.

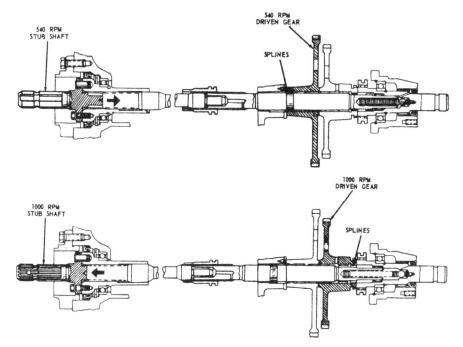

Fig. 166–Schematic view showing arrangement of pto shafts and gears used with dual 540-1000 rpm pto with mid-pto attachment. Note fore and aft movement of pto shafts.

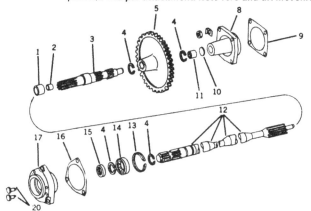

Fig. 167–Exploded view showing gears and shafts of the transmission driven 540 rpm pto.

1. Bushing
2. Bushing
3. Drive (front) shaft
4. Snap ring
5. Driven gear
8. Cover
9. Gasket
10. Thrust washer
11. Needle bearing
12. Power (rear) shaft
13. Snap ring
14. Ball bearing
15. Oil seal
16. Gasket
17. Bearing quill
20. Drive screw

REAR PTO SHAFT

All Models

160. **R&R AND OVERHAUL (540 RPM).** To remove rear pto shaft (12—Fig. 167), drain transmission and remove pto shield and shaft guard, if so equipped. Place rear pto shaft control lever in "OFF" position, then remove

1. Driven gear (540)
2. Driven gear (1000)
3. Spring
4. Ball
5. Drive screw
6. Detent spring
7. Ball
8. Dowel pin
9. Yoke
10. Shifter shaft
11. Coupling
12. Washer
13. "O" ring
14. Shifter lever
15. Retaining ring
16. Spring pins
17. Quill
18. Oil seal
19. Needle bearing
20. Snap ring
21. Ball bearing
22. Snap ring
23. Mid-stub shaft
24. Collar
25. Needle bearing
28. Quill
31. Rear stub shaft
32. "O" ring
33. Pilot
34. Oil seal
35. Snap ring
36. Washer, special
37. Ball bearing
38. Snap ring
39. Power (rear) shaft
40. Thrust washer
41. Bushing
42. Bushing
43. Drive (front) shaft
44. Thrust washer
45. Drive screw
48. Drive screw
49. Thrust washer
50. Cover
51. Gasket

cap screws from pto shaft bearing quill (17) and pull shaft and quill assembly from transmission case. Be careful not to pull shift collar from front drive shaft. Bearing (14) can be renewed after removing snap rings (4 and 13). Press new oil seal (15) in quill with lips toward front until it bottoms. For service on remainder of pto assembly, split tractor as outlined in paragraph 136.

Reassemble by reversing removal procedure and mate splines of rear shaft with splines of shift collar as shaft is installed.

161. **R&R AND OVERHAUL (540-1000 RPM).** To remove rear pto shaft (39—Fig. 168), drain transmission case and remove pto shield and shaft guard, if so equipped. Place rear pto shaft control in "OFF" position, then remove output (stub) shaft (31) from power-shaft pilot (33). Remove quill (28), then using caution not to damage bore, pry out oil seal (34). Remove snap ring (35), temporarily attach stub shaft (31) to function as a puller and remove power-shaft pilot (33) and washer (36). Power shaft (39) can now be removed but be careful not to pull rear pto shaft shift collar from front drive shaft.

Clean and inspect all parts. Bearing (37) can be removed from pilot after removing snap ring (38). Check condition of "O" ring (32).

To reassemble rear pto shaft, insert shaft (39) in rear of housing and turn shaft so shaft splines mate with shifter collar splines. Align splines of pilot (33) with splines of pto shaft, bump pilot and bearing assembly into place and install snap ring (35). Install oil seal (34) with lips toward front, then install quill (28), stub shaft (31) and shield and shaft guard, if so equipped.

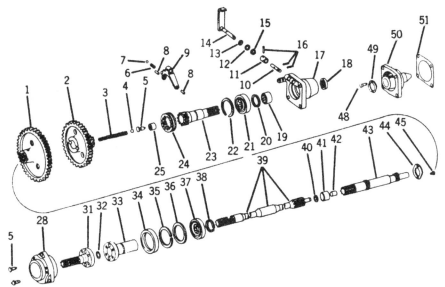

Fig. 168–Exploded view of dual 540-1000 rpm continuous-running pto shafts and gears. Items (6 through 24) are used with mid-pto units whereas items (48 through 51), are used on units with no mid-pto. Driven gears (1 and 2) are fitted with renewable bushings which are not shown.

DRIVEN GEARS, FRONT DRIVE SHAFT AND MID-PTO SHAFT

All Models

162. **R&R AND OVERHAUL.** To service the pto driven gears, front power shaft or mid-pto stub shaft, it is first necessary to split tractor as outlined in paragraph 136. Refer also to Fig. 169.

NOTE: The pto clutch power shaft and engine clutch shaft can also be removed at this time by withdrawing them rearward out of clutch housing. Pto drive gear can be pressed from pto clutch power shaft if necessary.

Prior to any disassembly, install rear pto shaft, if removed, and place rear pto shaft shifter lever in "ON" position. This will slide shift collar on rear pto shaft and prevent it from dropping to bottom of transmission case. If shift collar comes off, it will be necessary to remove the rockshaft housing to retrieve it.

Remove spring (3—Fig. 168) from front bore of drive shaft (43), remove driven gears (2 and 1), then pull drive shaft (43). See Fig. 170. Inspect thrust washer (T) at this time. Also inspect driven gear bushings and the drive shaft bushing in front bore of transmission. Renew as necessary.

To disassemble mid-pto stub shaft, remove ball (4—Fig. 168) from bore of stub shaft (23), then remove retaining ring (15), bump out roll pin and pull lever assembly from clutch housing. Bump out roll pin retaining shifter yoke (9), pull yoke shaft (10) from yoke, then remove yoke and shift collar (24) from stub shaft and catch detent ball (7) and spring (6) as yoke is removed. Remove snap ring (22) and bump stub shaft and bearing (21) rearward from quill (17). Straighten locks and remove quill stub nuts then bump quill from clutch housing.

NOTE: On models without mid-pto, only removal of cover (50) is involved and the procedure for doing so will be obvious.

Clean and inspect all parts. Bearing (21) can be pressed from stub shaft after removing snap ring (20). Needle bearing (25) and drive screw (5) located in bore of stub shaft are renewable. Oil seal (18) is installed in quill with metal side forward. Renew "O" ring (13) if doubt exists as to its condition.

Reassemble by reversing disassembly procedure and either use a piece of shim stock as a seal protector, or use caution, when inserting stub shaft through seal (18).

Fig. 169–View showing front end of transmission assembly. Rear gear (L) is 540 rpm driven gear. Front gear (H) is 1000 rpm driven gear.

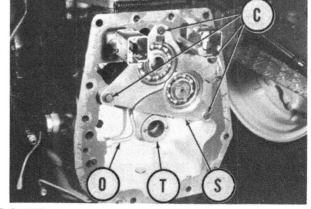

Fig. 170–View showing front of transmission case with pto drive shaft and both pto driven gears removed. Note thrust washer (T).

C. Cap screws
O. Oil line
S. Countershaft support
T. Thrust washer

REAR PTO SHAFT SHIFTER

All Models

163. **R&R AND OVERHAUL.** To remove rear pto shaft shifter assembly, remove the high and low range shifter shafts as outlined in paragraph 137 and the transmission countershaft as outlined in paragraph 138. Remove detent assembly located forward of and slightly below pto shift lever. Clip lock wire and remove shifter fork set screw.

Remove snap rings from ends of shifter shaft, push shifter shaft out front of transmission case and lift out shifter fork. Pull front pto shaft out and remove shift coupler. Bump roll pin from shifter arm and pull shifter lever from shifter arm and transmission case.

Clean and inspect all parts and renew as necessary. Reinstall shifter mechanism by reversing removal procedure.

INDEPENDENT POWER TAKE-OFF

Some tractors are optionally equipped with an independent power take-off which is driven by a splined hub in the single stage engine clutch cover. Control of the power shaft is entirely independent of transmission and is accomplished by a

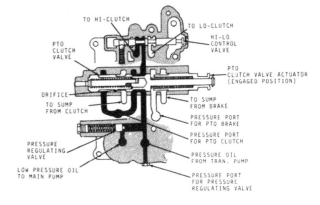

Fig. 171–Cross sectional view of iPTO control valve showing fluid flow. Refer to Fig. 176 for exploded view.

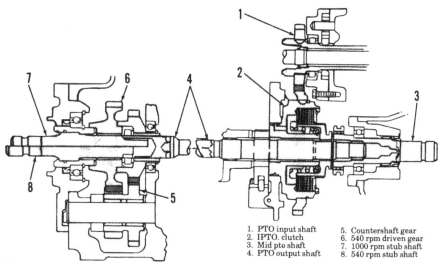

Fig. 172–Cross sectional view of IPTO power train used in 540-1000 rpm models. The 540 rpm unit is similar except that reduction gears (5 & 6) and stub shafts (7 & 8) are not used and shaft (4) extends rearward out of transmission housing. Clutch and front reduction ratio also differ.

1. PTO input shaft
2. IPTO. clutch
3. Mid pto shaft
4. PTO output shaft
5. Countershaft gear
6. 540 rpm driven gear
7. 1000 rpm stub shaft
8. 540 rpm stub shaft

hydraulically engaged multiple disc clutch contained in clutch housing.

Independent Power Take-Off is available in combination with, or without, Hi-Lo Shift but cannot be used with Reverser unit.

OPERATION

Models With Independent Power Take-Off

164. Refer to Fig. 171 for cross sectional view of control valve and to Fig. 172 for cross section of power train.

Fig. 173–View of shift cover showing IPTO lever installed.

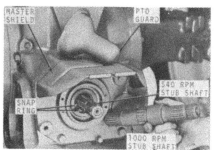

Fig. 174–Stub shafts are interchangeable from rear without draining transmission.

Hydraulic power to engage the pto clutch and brake is provided by the transmission pump and is combined with the Hi-Lo clutch circuit on models so equipped.

The pto control valve is designed so there is no overlap between clutch and brake circuits. When control lever is moved to the engaged position, pressure fluid flows to the pto clutch and at the same time enters the area behind

Fig. 175–View of shift cover with test port plugs identified.

Fig. 176–Exploded view of shift cover and valves on models equipped with IPTO and Hi-Lo Shift Unit.

1. Hi-Lo Clutch spool
2. Shift cover
3. Regulating valve
4. Test plug
5. Detent assembly
6. Detent pin
7. Cover
8. IPTO shift valve

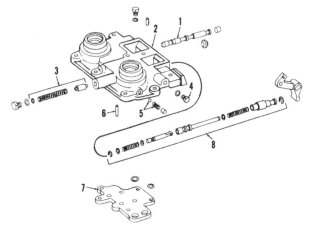

the valve, preventing rapid opening of the valve and consequent rough clutch engagement. At the same time pressure is ported to the clutch, pto brake pressure is released and brake piston passage opened to the sump.

When control lever is moved to the disengaged position, clutch passage is opened to the sump and pressure applied to a piston located in countershaft bearing support which applies a band-type brake to pto output shaft.

TESTING

Models With Independent Power Take-Off

165. Test ports for brake and clutch passages are shown in Fig. 175, or circuit pressure can be checked by removing plug (4—Fig. 176). Pressure should be 140-160 psi at any port when that circuit is open, checked at operating temperature and 2100 engine rpm.

If regulated pressure cannot be properly adjusted, refer to paragraph 171 for additional hydraulic system checks and to paragraphs 166 through 169 for overhaul procedure.

OVERHAUL

Models With Independent Power Take-Off

166. **CONTROL AND PRESSURE REGULATING VALVES.** Control and pressure regulating valves are located in shifter cover as shown in Fig. 176. Regulating valve can be adjusted without removing shifter cover, but overhaul of valve can only be accomplished after cover is removed; proceed as follows:

Remove the four cap screws retaining transmission shield and remove the shield by working it up over shift lever boots. Disconnect Hi-Lo linkage if so equipped. Disconnect rear wiring harness. Shift IPTO control lever to engaged position, then remove lever and housing. Remove remaining cap screws and lift off shift cover, valves and shift levers.

Pin (6) is inserted behind spring of valve detent (5). Remove plate (7) and withdraw the pin before attempting to remove Hi-Lo valve spool (1). Valve spool (1), detent (5) and pin (6) will not be present in models not equipped with Hi-Lo shift unit.

Clean and inspect all parts. Front and rear springs in IPTO shift valve (8) are interchangeable and three spacer washers are normally used in each spring location. Rear spring should require a pressure of approximately 7 lbs. to move seating washer away from head of guide pin. Add or remove washers to adjust, provided at least one washer is used. (Three is normal). With valve, front spring and actuator assembled, a pressure of approximately 13 lbs. should be required to move actuator away from front snap ring. Add or remove washers behind front spring to adjust.

Fig. 177 shows an exploded view of IPTO valve lever and associated parts. The two detent screws (2) should be tightened until they bottom (springs solidly compressed); then backed out two full turns to apply the required detent pressure.

167. IPTO CLUTCH AND INPUT GEARS. Refer to Fig. 178 for an exploded view of clutch assembly. To remove the clutch, first split tractor between clutch housing and transmission case as outlined in paragraph 136.

On units equipped with mid-pto, shift rear selector lever to engaged position to prevent the disconnect collar from falling into transmission case and withdraw clutch unit and pto drive shaft as an assembly. On units without mid-pto, remove front snap ring (18) and withdraw clutch unit only, leaving pto drive shaft in transmission housing.

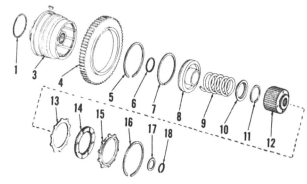

Fig. 178–Exploded view of IPTO clutch and associated parts. Refer also to paragraph 167 for parts identification on various models.

1. Sealing ring	7. Piston ring
3. Housing	8. Piston
4. Drive gear	9. Spring
5. Snap ring	10. Retainer
6. Sealing ring	

11. Snap ring	15. Backing plate
12. Clutch hub	16. Snap ring
13. Separator plates	17. Thrust washer
14. Clutch discs	18. Snap ring

Clutch housings (3) differ on units equipped with 540 rpm single speed pto and 540-1000 rpm models. Gear (4) also differs, the 540 rpm unit having 73 teeth and the dual speed unit 62 teeth. On 540 rpm units the clutch turns at 540 rpm at rated speed and higher torque is transmitted. Eight clutch discs (14) and separator plates (13) are used. On dual speed units the clutch turns at 1000 rpm and speed reduction takes place at output end. Four each clutch discs (14) and separator plates (13) are used on dual speed units. Drive gear (4) is installed on clutch drum with offset forward on models equipped

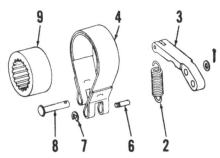

Fig. 179–Exploded view of IPTO shaft brake and associated parts. Refer to Fig. 180 for assembled view and for actuating piston and cylinder assembly.

2. Spring	7. Retainer
3. Lever	8. Anchor pin
4. Band	9. Brake drum
6. Pivot pin	

with Hi-Lo Shift; or offset rearward on other models. Check the gear before removal, for proper installation.

Input gear can be withdrawn from rear of clutch housing on models without Hi-Lo shift, or removed with oil manifold (or transmission oil pump) on models with Hi-Lo shift.

To remove clutch piston (8) compress spring (9) using a press and suitable fixture, then remove snap ring (11). Gear (4) should be heated to approximately 360° before pressing on pto clutch drum.

168. IPTO BRAKE. The band type IPTO brake mounts on transmission countershaft bearing support and acts on output shaft to stop the pto shaft when clutch is disengaged and brake applied. Refer to Figs. 179 and 180 for exploded views.

To remove the IPTO brake, first remove clutch as outlined in paragraph 167; then remove countershaft and bearing support as in paragraph 138. Brake band is mounted on bearing support as shown, and actuating piston is carried in bearing support cylinder bore. Renew brake band if total thickness (band & lining) is ⅛ inch or less.

169. OUTPUT SHAFT AND GEARS. Refer to Fig. 181 for an exploded view of dual IPTO shafts and to

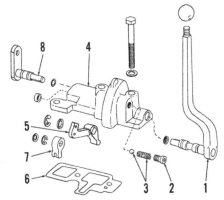

Fig. 177–Exploded view of IPTO valve lever and associated parts.

1. IPTO shift lever	5. Lever arm
2. Detent plug	6. Gasket
3. Detent assembly	7. Hi-Lo Shift valve arm
4. Cover	8. Valve shaft

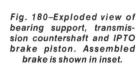

Fig. 180–Exploded view of bearing support, transmission countershaft and IPTO brake piston. Assembled brake is shown in inset.

6. Pivot pin	
8. Anchor pin	
10. Piston	
11. Plug	
12. Bearing support	
13. Transmission drive gear	
14. Countershaft	
15. Friction plug	

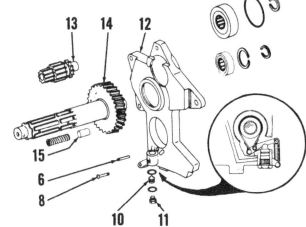

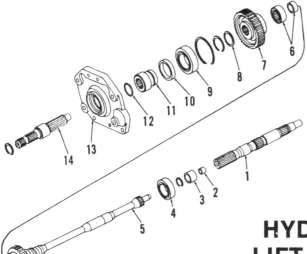

Fig. 181—Exploded view of IPTO drive gears and shaft units showing component parts.

1. IPTO drive shaft
2. Bushing
3. Bushing
4. Bearing
5. IPTO output shaft
6. Bearing
7. 540 rpm gear
8. Spring washer
9. Bearing
10. Oil seal
11. Shaft pilot
12. O-ring
13. Bearing quill
14. Stub shaft

HYDRAULIC LIFT SYSTEM

Fig. 182 for cross section of reduction gears and associated parts.

To remove the rear pto shaft, reduction gears and associated parts, drain transmission and remove pto shield and shaft guard. Move rear pto lever to "OFF" position. Remove stub shaft, if installed. Remove the cap screws retaining pto shaft bearing quill, remove the quill then withdraw gears, shafts and associated parts. When installing bearing quill on dual IPTO models, adjust draft control negative stop screw as outlined in paragraph 175.

On models with 540 rpm single speed IPTO, shaft and rear bearing quill will be removed as a unit. On models without mid-pto, it will be necessary to remove rockshaft housing as outlined in paragraph 182 and install coupling sleeve between pto output shaft and pto drive shaft.

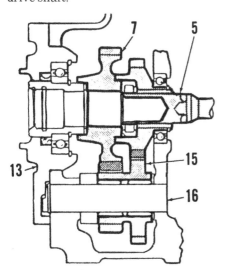

Fig. 182—Cross sectional view of IPTO reduction gears and associated parts. Refer to Fig. 181 for parts identification except for the following:

15. Countershaft gear
16. Counter shaft

The hydraulic lift system is a closed center, constant pressure type. The stand-by 2200-2300 psi pressure is furnished by either a four or eight piston, constant running, variable displacement pump which is mounted in the tractor front support and is driven by a coupling from front end of engine crankshaft. Charging oil for the hydraulic main pump is supplied by the transmission oil pump and oil not used by the main hydraulic pump is routed to the auxiliary hydraulic oil reservoir which provides an auxiliary supply of oil when transmission oil pump is unable to meet the demand of main hydraulic pump. When there is little or no demand by the main hydraulic oil pump, the overflow from auxiliary reservoir is returned to the clutch housing (through an oil cooler on some models) where part of it fills the brake control valve reservoir and the remainder lubricates the transmission shafts and gears.

TROUBLE SHOOTING

All Models

170. The following are symptoms which may occur during the operation of the hydraulic lift system. By using this information in conjunction with the Test and Adjust information, no trouble should be encountered in servicing the hydraulic system.

1. Slow system operation. Could be caused by:
   a. Clogged transmission oil filter.
   b. Transmission oil pump inlet screen plugged.
   c. Faulty transmission oil pump.
   d. Transmission oil pump relief stuck open.
   e. Hydraulic pump stroke control valve not seating properly.
   f. Hydraulic pump crankcase out-filter plugged.
   g. Oil leak on low pressure side of system.

2. Erratic pump operation. Could be caused by:
   a. Pump stroke control valve not seating properly.
   b. Leaking pump inlet or outlet valves or valve "O" rings.
   c. Broken or weak pump piston springs.

3. Noisy pump. Could be caused by:
   a. Worn drive parts or loose cap screws in drive coupling.
   b. Air trapped in oil cavity of pump stroke control valve.

4. No hydraulic pressure. Could be caused by:
   a. Pump shut-off valve closed (if so equipped).
   b. No oil in system.
   c. Faulty pump.

5. Rockshaft fails to raise or raises slowly. Could be caused by:
   a. Excessive load.
   b. Low pump pressure or flow.
   c. Rockshaft piston "O" ring failed.
   d. Flow control valve maladjusted.
   e. Thermal relief valve defective.
   f. Cam follower adjusting screw maladjusted.
   g. Transmission oil filter plugged.
   h. Defective seals between cylinder and rockshaft housing, or between rockshaft housing and transmission case.

6. Rockshaft settles under load. Could be caused by:
   a. Leaking discharge valve.
   b. Leaking rockshaft cylinder check valve.
   c. Leaking cylinder pipe plug.
   d. Faulty rockshaft cylinder valve housing.

7. Rockshaft valves hunt. Could be caused by:
   a. Control valves maladjusted.
   b. Rockshaft piston "O" ring faulty.
   c. Discharge valve leaking.
   d. Thermal relief valve leaking.

8. Rockshaft lowers too fast or too slow. Could be caused by:
   a. Throttle valve maladjusted.
   b. Broken or disconnected valve linkage.

9. Rockshaft raises too fast. Could be caused by:
   a. Flow control valve incorrectly set.

10. Insufficient load response. Could be caused by:
    a. Control valve clearance excessive.
    b. Control valves sticking.
    c. Control lever not positioned correctly on quadrant.
    d. Worn load control shaft or bushings.

e. Negative stop screw turned too far in.
11. Hydraulic oil too hot. Could be caused by:
   a. Control valves adjusted too tight and held open.
   b. Control valves leaking.
   c. Control valve "O" rings faulty.
   d. Thermal relief valve faulty (leaking).

HYDRAULIC SYSTEM TESTS

All Models

Before making any tests on the main hydraulic pump or lift system, be sure the transmission oil pump is satisfactory as the performance of main hydraulic pump is dependent upon being charged by the transmission oil pump. For information on testing of transmission oil pump, refer to paragraph 142.

171. **MAIN HYDRAULIC PUMP TEST.** The main hydraulic pump can be tested for standby pressure and flow rate by using a hydraulic test unit. Pump stand-by pressure can also be checked by using a 3000 psi gage. Refer to paragraph 172 for procedure using hydraulic test unit and to paragraph 173 for procedure using pressure gage

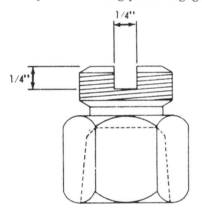

Fig. 183–Modify adapter as shown when attaching hydraulic tester outlet line. Refer to text.

only. Note: On models equipped with selective remote control valves, main hydraulic pump pressure can also be checked as outlined in paragraph 186.

172. When testing pump using the hydraulic test unit, proceed as follows: Disconnect pump pressure line from tee (remote valves) or elbow (no remote valves) located on lower right side of transmission case. Use necessary fittings and connect inlet line of test unit to open end of pump pressure line. Remove the large hex head plug directly above the transmission oil filter. Obtain a special adapter, John Deere R1961R or equivalent (1 1/16 to ¾), modify it as shown in Fig. 183 so it will not interfere with filter relief valve, then install it in the removed plug hole and connect outlet line of test unit. See Fig. 184.

NOTE: If flow rate of main hydraulic pump is to be checked, it is necessary to connect hydraulic test unit as outlined above so that discharged oil will be directed back to the pump inlet to insure an adequate oil supply for the main hydraulic pump.

Start engine and run at 2500 rpm. Close the hydraulic test unit control valve and note the pressure gage which should read 2200-2300 psi. This is the pump stand-by pressure. If pump stand-by pressure is not as stated, loosen jam nut and turn pump stroke control valve adjusting screw (Fig. 185) in to increase pressure or out to decrease pressure.

With the correct pump stand-by pressure established, set tester control valve so tester pressure gage reads 2000 psi which is the system working pressure. At this pressure, the tester flow meter should show a 6 gpm minimum flow for the four piston pump, or 11 gpm minimum flow for the eight

piston pump. If main hydraulic pump will not meet both of the above conditions it must be removed and overhauled as outlined in paragraph 180.

173. When no hydraulic test unit is available, test main hydraulic pump stand-by pressure as follows: Disconnect the pump pressure line from the tee, or elbow, located on right side of transmission, then install a tester which includes a 3000 psi gage and a shut-off valve in series with the pressure line and be sure the gage is ahead of the shut-off valve. Start engine and run at 2500 rpm. Close the shut-off valve and note the gage reading which should be 2200-2300 psi. This is the main hydraulic pump stand-by pressure. If pump stand-by pressure is not as stated, loosen jam nut and turn pump stroke control valve adjusting screw (Fig. 185) in to increase pressure or out to decrease pressure. If pump will not produce the 2200-2300 psi pressure it must be removed and overhauled as outlined in paragraph 180.

174. **FLOW CONTROL VALVE TEST.** When tractors are equipped with an eight piston pump, a flow control valve is incorporated into the rockshaft valve circuit to reduce the oil flow. See Fig. 186. Check the flow control valve setting with a hydraulic test unit as follows: Connect the hydraulic

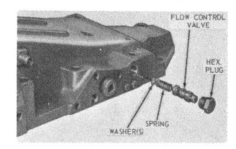

Fig. 186–When tractor is equipped with an eight piston pump a flow control valve is installed in right side of rockshaft housing.

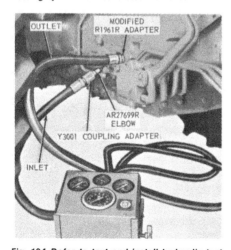

Fig. 184–Refer to text and install hydraulic test unit as shown. Refer also to Fig. 183.

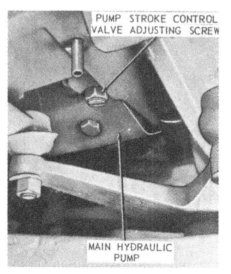

Fig. 185–Hydraulic pump stroke control valve is located as shown.

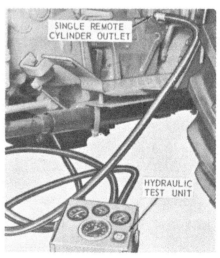

Fig. 187–When testing flow control valve, refer to text and connect hydraulic tester unit as shown.

test unit outlet line as outlined in paragraph 172, then remove the single remote cylinder outlet plug located on left side of rockshaft housing and connect the hydraulic test unit inlet line to plug hole. See Fig. 187. Place the rockshaft load control selector lever in "L" position, start engine and run at 2500 rpm, then move the rockshaft control lever to maximum lift position. Close the tester control valve until tester pressure gage shows 1000 psi, then check flow meter gage which should show a 4¾-5¾ gpm flow. If oil flow is not as stated, refer to Fig. 186, and remove plug, valve and spring, then vary washers (shims) as required. Also check spring which should test 7½-9 lbs. when compressed to a length of 53/64-inch.

If tractor is equipped with selective (remote) control valves, refer to paragraph 185 for testing information.

HYDRAULIC SYSTEM ADJUSTMENTS

All Models

The following paragraphs outline the adjustments that can be made when nec-

Fig. 188–Negative stop adjusting screw is located on right rear of transmission case.

Fig. 189–Arrow shows negative stop screw on models with dual IPTO.

essary to correct faulty hydraulic operation, or that must be made when reassembling a hydraulic lift system that has been disassembled for service.

However, because of the interaction of component parts, the adjustments must be made in the following order.

A. Negative stop screw
B. Rockshaft control lever neutral range
C. Control lever position
D. Load control
E. Rate-of-drop

175. **NEGATIVE STOP SCREW ADJUSTMENT.** The negative stop screw is located on the right rear of the transmission case as shown in Fig. 188, on models without dual independent power take-off; or as shown at arrow— Fig. 189 on models with dual IPTO.

To adjust the negative stop screw, loosen jam nut and turn stop screw in until it just contacts the load control arm, then turn (back) screw out 1/6-turn and tighten jam nut.

NOTE: Contact of stop screw with load control arm can be more easily felt if rockshaft housing filler plug is removed and a screw driver is held against upper end of the load control arm.

176. **CONTROL LEVER NEUTRAL RANGE ADJUSTMENT.** To adjust the control lever neutral range, first remove the pipe plug, located directly in front of control lever tube which will expose the control valve adjusting screw. See Fig. 190. Place selector lever in upper (D) notch, start engine and run at 2500 rpm. Start with rockshaft control lever at rear of quadrant, move lever slowly forward until rockshaft just starts to lower and mark this point on upper edge of quadrant. Now move lever slowly rearward until rockshaft just begins to raise and mark this point on upper edge of quadrant.

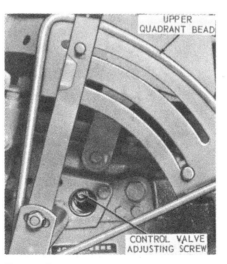

Fig. 190–Control valve adjusting screw is located under a plug which is in front of control shaft tube.

Distance between these two marks should be 1/16-3/16 inch.

If neutral range of rockshaft control lever is not as stated, turn the control valve adjusting screw clockwise to increase, or counterclockwise to decrease, the control lever neutral range.

177. **CONTROL LEVER POSITION ADJUSTMENT.** Place load selector in upper (D) position, start engine and run at 2500 rpm, then move rockshaft control lever fully forward to completely lower rockshaft.

On RU and HU model tractors, loosen rockshaft control lever adjusting nut (Fig. 191) and move control lever rearward until there is 5/16-inch clearance between control lever friction pin and bottom end of quadrant slots.

On LU model tractors, move rockshaft control lever fully forward to completely lower rockshaft, disconnect swivel (Fig. 192) from control shaft arm, then move control lever rearward until there is 3/16-inch clearance between control lever friction pin and bottom end of quadrant slot.

On all models, rotate control shaft arm clockwise as far as possible, then

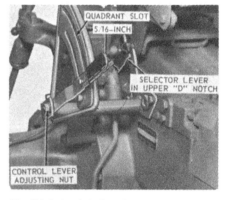

Fig. 191–Set rockshaft control lever as shown for models RU and HU when making control lever adjustment.

Fig. 192–Set rockshaft control lever as shown for model LU when making control lever adjustment.

rotate control shaft arm counter-clockwise to the point where rockshaft just starts to raise. On RU and HU models, tighten control lever adjusting nut. On LU models, adjust swivel until it enters hole in control shaft arm without moving arm.

178. **LOAD CONTROL ARM ADJUSTMENT.** Remove the rockshaft filler hole plug to expose the load control arm cam follower adjusting screw. See Fig. 193. Place selector lever in lower (L) position, start engine and run at 2500 rpm. Move rockshaft control lever fully forward to completely lower rockshaft, then move control lever rearward until distance between control lever friction pin and top end of quadrant slot is 1⅞ inches for RU and HU model tractors, or 59/64-inch for LU model tractors. See Fig. 194. Hold rockshaft control lever in this position, loosen jam nut and turn cam follower adjusting screw counter-clockwise until rockshaft is fully lowered, then turn adjusting screw clockwise until rockshaft starts to raise. Tighten jam nut.

NOTE: If rockshaft begins to raise before rockshaft control lever reaches the 1⅞ or 59/64-inch setting, it will be necessary to turn cam follower adjusting screw counter-clockwise to allow the lever to be positioned while rockshaft remains in lowered position.

179. **RATE-OF-DROP ADJUST-MENT.** The rate of drop (throttle) adjusting screw is located on top side of rockshaft housing as shown in Fig. 195. Turn adjusting screw clockwise to decrease rate-of-drop, or counter-clockwise to increase rate of drop. Tighten jam nut after adjustment is made. Rate of drop will vary with the weight of the attached implement.

MAIN HYDRAULIC PUMP

All Models

The main hydraulic pump may be either a four piston or an eight piston type having an output of 6½ gpm or 13 gpm respectively at 2500 engine rpm. Pumps are identical in operation and construction, except for the number of pistons and valves.

Pumps maintained a 2000 psi working pressure with a stand-by pres-

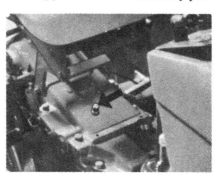

Fig. 195—View showing location of the rate-of-drop adjusting screw.

sure of 2200-2300 psi. Stand-by operation of the pump occurs when pressure in the pump crankcase builds to 2200-2300 psi thereby holding the pump pistons away from the pump cam. The pump crankcase (stand-by) pressure is controlled by stroke control valve located in a bore in the pump housing.

A pump shut-off (destroking) screw (35—Fig. 196) is optionally available which will make pump inoperative and will act as an aid during cold weather starts.

180. **R&R AND OVERHAUL.** The hydraulic pump can be removed without radiator being removed, however, removal of radiator provides additional working room. The following procedure is based on radiator not being removed.

To remove hydraulic pump, first remove hood and side panels. Drain radiator and remove lower hose. Disconnect reservoir to pump line and drain reservoir. Disconnect inlet and pressure lines from pump. Remove air cleaner assembly. Remove engine timing hole plug and turn engine until the clamping cap screw at forward end of pump drive coupling is at about the 10 o'clock position. Use a long extension and remove cap screw. Remove pump mounting bolts, pull pump forward to remove pump shaft from drive

Fig. 193— Cam follower adjusting screw is exposed when filler plug is removed.

Fig. 194—Position control lever as shown in left view for RU and HU models, or as shown in right view for LU models, when making cam follower adjustment.

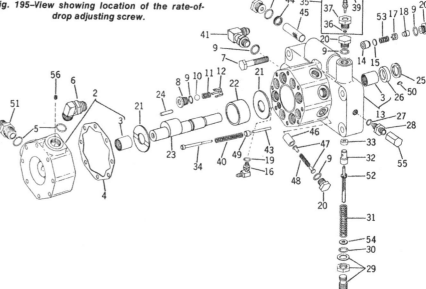

Fig. 196—Exploded view of 8 piston main hydraulic pump. Except for the number of pistons (46), inlet valves (10) and outlet valves (15) the 4 piston pumps are identical.

2. Cover	17. Guide	29. Adjusting screw assy.	43. Pin
3. Needle bearing	18. Stop	30. "O" ring	44. Seal (2 used)
4. Gasket	19. "O" ring	31. Spring	45. Filter screen
5. "O" ring	20. Plug	32. Stroke control valve	46. Pump piston
6. Adjustable elbow	21. Thrust washer	33. Valve seat	47. Guides
8. Inlet valve seat	22. Cam race	34. Spring guide	48. Piston spring
9. "O" ring	23. Pump shaft (cam)	35. Pump shut-off assy.	49. Crankcase outlet
10. Inlet valve ball	24. Roller bearing (33	36. "O" ring	valve
11. Spring	used)	37. Gland nut	50. Woodruff key
12. Guide	25. Quad ring	38. Shut-off screw	51. Connector
13. Housing	26. Oil seal	39. Spring pin	52. Valve guide
14. Outlet valve seat	27. "O" ring	40. Spring	53. Spring
15. Outlet valve	28. Connector	41. Adjustable elbow	55. Cap
16. Adjustable elbow		42. Plug	56. Orifice

coupling, then disconnect bleed line and remove pump from left side of tractor.

With pump removed, clean unit and prior to any disassembly, use a dial indicator and check pump shaft end play. Shaft end play should be 0.004-0.038 and if end play is excessive, renew thrust washers (21—Fig. 196) when reassembling. Clamp pump in a vise and remove cover (2), then remove crankcase outlet valve (items 34, 40, 43 and 49—Fig. 196) from pump housing. Also see Fig. 197. Remove outer thrust washer (21—Fig. 196).

NOTE: At this time, shaft (23) can be removed if desired by removing woodruff key (50) and pushing shaft out of body and cam ring (22). Be sure not to lose any of the 33 rollers (24) which will be loose. However, if pump requires disassembly, it is good policy to completely disassemble pump and inspect all parts and to remove pump pistons before removing pump shaft. See Fig. 198.

Remove pistons assemblies (20, 46, 47 & 48—Fig. 196), then if not already removed, remove shaft (23), rollers (24) and cam ring (22). Remove inlet valves (8, 10, 11 and 12) and outlet valves (11, 15, 17, 18 and 20). Identify all pistons, valves, springs and seats so they can be reinstalled in their original positions. Remove plug (42) and remove filter screen (45). Also see Fig. 199. Loosen jam nut and remove stroke control valve adjusting screw (29), spring (31), spring guide and valve (32). Also see Fig. 200.

Clean and inspect all parts. Use data in paragraph 181 as a guide for renewal of parts and assembly. Discard all "O" rings and use new during assembly. On some early pumps, split bushings are used instead of shaft needle bearings (3—Fig. 196). Bushings are renewable and are installed flush with inner ends of bushing bores. Pay particular attention to shaft bore around the quad ring seal groove as leakage at this point can cause the pump to be slow in going out of stroke. If outlet valve seats (14) are damaged, drive them out and install new seats with large chamfered end toward bottom of bore. Press seats in to 1.171 below spot face surface of bore. O.D. of new outlet valve (15) is 0.609-0.611 and any valve that is distorted, scored or worn should be renewed. If stroke control valve seat (33) is damaged, remove plug (20), or shut-off assembly (35) if so equipped, and drive out seat (33). Install new valve chamfered end first and drive it in bore until it bottoms. Spring (31) should test 158-192 lbs. when compressed to a length of 2½ inches. Renew any pistons (46) which are scored or pitted. No specific test value is given for piston springs (48), however springs MUST test within 0.15 lb. of each other when compressed to a length of 1¼ inches.

When reassembling, dip all parts in oil. Use grease to hold rollers in I.D. of cam ring during assembly. Seal (26) is installed with printed side toward outside. Thrust washers are installed with grooved sides away from pump shaft cam.

181. The following data applies to all pumps and can be used as a guide for parts renewal and assembly.

Thrust washer thickness . . 0.087-0.091
Shaft bushing bore I.D.
(early models) 1.1245-1.1255
Piston bore I.D. 0.6802-0.6808
Pump shaft cam O.D. 1.610-1.670
Cam race O.D. 2.235-2.245
Cam race I.D. 1.8004-1.8010
Crankcase bore depth 2.660-2.666
Torque, cover to body screws 36 ft.-lbs.
Torque, piston plugs . . . . 90 ft.-lbs.
Torque, pump mounting
screws . . . . 85 ft.-lbs.

ROCKSHAFT HOUSING, CYLINDER AND VALVE

All Models

182. **REMOVE AND REINSTALL.** To remove rockshaft housing, disconnect battery ground straps and on RU and HU models, remove transmission shield and disconnect wires from starter safety switch. On LU models, disconnect swivel from control shaft arm and remove quadrant and rod along with shield. Disconnect lift links from rockshaft arms. Remove seat assembly. If tractor is equipped with remote hydraulic system, disconnect lines from rear of selective (remote) control valves and remove return hose between valve and rockshaft housing. Attach hoist to rockshaft housing, place selector lever in lower (L) posi-

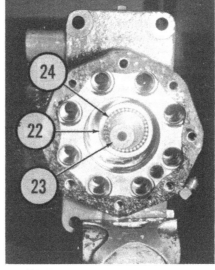

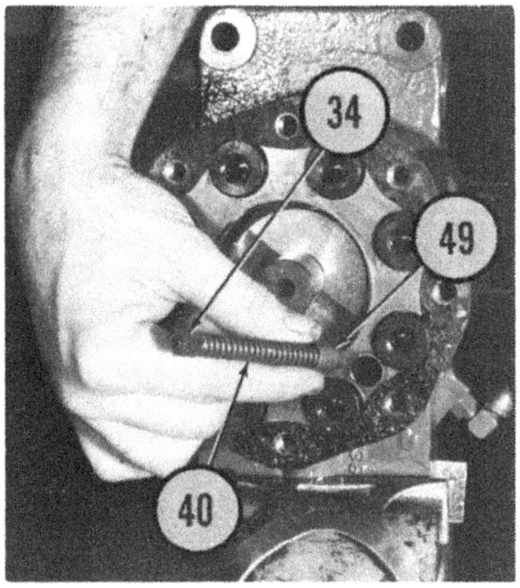

Fig. 198–View showing pump with cover, inlet valves and pistons removed. Pump shaft (23), rollers (24) and cam race (22) can be removed as a unit, if desired.

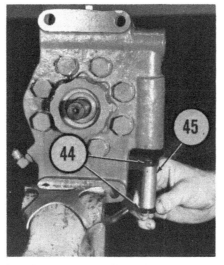

Fig. 197–Crankcase outlet valve is removed from bore as shown. Valve pin is still in bore.

Fig. 199–Remove pump filter (45) as shown. Two seals (44) are used.

Fig. 200–Pump stroke control valve being removed. Valve guide (52–Fig. 196) is in inside spring (31).

tion, then unbolt and lift rockshaft housing from transmission case. See Fig. 201.

Installation is the reverse of removal but be sure dowels are in place and that roller link mates properly with the load control arm cam follower.

183. **OVERHAUL.** With rockshaft housing assembly removed, disassemble as follows: Remove rockshaft arms from rockshaft. Remove selector lever from load control arm. Turn unit so bottom side is accessible, then drive out spring pin (SP—Fig. 201) which retains control shaft to pivot block.

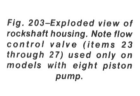

Fig. 203–Exploded view of rockshaft housing. Note flow control valve (items 23 through 27) used only on models with eight piston pump.

Fig. 201–View showing bottom side of rockshaft housing. Spring pin (SP) retains control lever shaft to pivot block.

2. Oil filler cap	11. Remote cylinder outlet
3. Gasket	adapter
4. Retainer	12. Gasket
5. "O" ring	16. Aluminum washer
6. Bushing	17. Starter safety switch
8. Pin	18. "O" ring
9. Plug	19. Throttle valve adjusting screw
10. "O" ring	20. Jam nut
	21. Plug

22. "O" ring	31. Nut
23. Washer (shim .048)	32. "O" ring
24. Spring	33. Bushing
25. Flow control valve	34. Plug
26. Plug	35. "O" ring
27. "O" ring	36. Load control arm
28. Plug	37. Gasket
29. Load selector lever	38. Rockshaft housing
30. Washer	

Unbolt plate from rockshaft housing remove quadrant and control shaft (RU and HU) or plate and control shaft (LU). Remove remote cylinder outlet adapter from left side of housing and unhook valve spring from linkage. Remove front outside cylinder mounting cap screw, then remove remaining cylinder mounting cap screws, lift cylinder assembly from housing and disengage selector arm from roller link as cylinder is removed. Do not lose throttle valve ball as it will fall free.

Remove cam from rockshaft, then pull rockshaft from control arm and housing. Bushing, "O" ring and retainer will be removed by the rockshaft on the side rockshaft is removed from.

If necessary, separate piston rod from control arm by driving out spring pin. Load selector shaft can also be removed if necessary. On tractors equipped with an eight piston hydraulic pump, remove flow control valve assembly from right front of hydraulic housing.

With components removed, disassemble cylinder and valve unit as follows: Remove plug (10—Fig. JD-202), spring (12) and check ball (2). Remove thermal relief valve (14). Remove retainer ring and pull linkage from linkage pivot pin. Remove snap rings (25), valve plugs (23), spring (20), sleeves (29), valves (19) and seats (18) from bores in cylinder housing. Keep valve assemblies identified so they can be reinstalled in original bores. Remove piston from cylinder by bumping open end of cylinder against a wood block or remove plug (17) and push piston out. Remove and discard piston "O" ring and back-up ring.

Clean and inspect all parts for undue wear, bends or other damage. Use Figs. 202, 203, 204 and 205 to identify parts. Check valve spring (12—Fig. 202) should test 22½-27½ lbs. when compressed to a length of 2⅛ inches. Con-

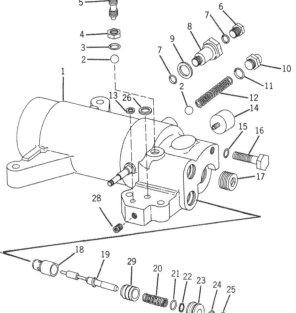

Fig. 202–Exploded view showing control valve components. Only one valve assembly is shown, however other valve is identical.

1. Cylinder & valve hsg.
2. Steel ball
3. "O" ring
4. Jam nut
5. Throttle valve adjusting screw
6. Plug
7. "O" ring
8. Remote cylinder outlet adapter
9. Gasket
10. Plug
11. "O" ring
12. Spring
13. Gasket
14. Thermal relief valve
15. "O" ring
16. Cap screw
17. Plug
18. Valve seat
19. Control valve
20. Spring
21. "O" ring
22. Back-up washer
23. Plug
24. "O" ring
25. Snap ring
26. Seal
28. Plug
29. Sleeve

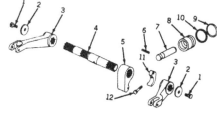

Fig. 204–View of rockshaft, piston and lift arms.

1. Cap screws	7. Piston rod
2. Washer	8. Piston
3. Lift arm	9. "O" ring
4. Rockshaft	10. Back-up ring
5. Control arm	11. Cam
6. Spring pin	12. Cap screw

trol valve springs (20) should test 8 5/16 lbs. when compressed to a length of ⅞-inch. New piston diameter is 3.244-3.246. If so equipped, flow control valve spring (24—Fig. 203) should check 7½-9 lbs. when compressed to a length of 53/64-inch. Pay particular attention to seating area of the two control valves (19—Fig. 202). Leakage in this area could cause rockshaft settling if it occurs in the discharge (upper) valve, or upward rockshaft creep if it occurs in the pressure (lower) valve. Also pay close attention to the valve seats (18) and sleeves (29). Inspect rockshaft assembly for fractures, damaged splines or other damage. Check control linkage for worn or bent condition. A worn adjusting cam (10—Fig. 205) will cause difficulty in adjusting lever neutral range.

When reassembling the rockshaft and cylinder unit, use sealant on threads of any pipe plugs that were removed and coat all "O" rings with oil.

Start rockshaft into rockshaft housing, align master splines of rockshaft and control arm, then push rockshaft into position. Install rockshaft bushings, which will be free in rockshaft bores, then install "O" rings and retainers with cupped side outward. Install cam on rockshaft and connecting rod to control arm. Insert valve seats (18—Fig. 202) in their original bores small end first, then place valves (19) in seats with smaller diameter of valves in seats. Install sleeves (29), chamfered end first, and valve springs (20). Install "O" rings (21) and back-up rings (22) in I.D. of plugs (23) with back-up rings toward outer end of valves. Install "O" rings (24) on O.D. of plugs, then carefully install plugs over ends of valves and install snap rings (25). Install "O" ring (9—Fig. 204) and back-up ring (10) on piston (8) with "O" ring toward closed end of piston. Lubricate piston assembly and install it in cylinder. Install check valve and thermal relief valve. Assemble linkage and install it on linkage pivot pin. If unit is so equipped, install flow control valve. If not already done, remove the throttling valve screw (5—Fig. 202)

from rockshaft housing. Install selector control lever shaft in rockshaft housing. Place "O" ring (15) in front cap screw hole of cylinder and valve unit and the gasket (13) and seal (26) on flange of valve housing, then install cylinder and valve assembly in rockshaft housing and start slot of roller link over selector control lever shaft and enter connecting rod in piston as cylinder valve unit is positioned. Install and tighten the front cap screw (16) before tightening the other cylinder mounting cap screws. Connect the linkage spring and install the throttling valve ball (2) and adjusting screw (5). Install control quadrant assembly and selector lever.

Install rockshaft housing assembly by reversing the removal procedure and adjust assembly as outlined in paragraphs 175 through 179.

LOAD CONTROL (SENSING) SYSTEM

All Models

The load sensing mechanism is located in the rear of the transmission case. See Fig. 206 for an exploded view showing component parts.

The load control shaft (21) is mounted in tapered bushings and as load is applied to the shaft ends from the hitch draft links, the shaft flexes forward and actuates the load control arm (12) which pivots on shaft (10). Movement of the load control arm is transmitted to the rockshaft control valves via the roller link (2—Fig. 205)

and control linkage and control valves are opened or closed permitting oil to flow to or from the rockshaft cylinder and piston.

184. **R&R AND OVERHAUL.** To remove the load sensing mechanism, remove the rockshaft housing assembly as outlined in paragraph 182, the left final drive assembly as outlined in paragraph 150, and the three-point hitch.

Remove cam follower spring (18—Fig. 206), then slide pivot shaft (10) to the left and lift out control arm (12) assembly. Removal of pivot shaft (10) can be completed if necessary by removing snap ring (11). Remove retainer ring (2) and retainer bushing (3) from right end of load control shaft and bump shaft from transmission case. The negative stop screw (7), located on rear of transmission case behind right final drive, can also be removed if necessary.

Inspect bushings (6) in transmission case and renew if necessary. Drive old bushings out by inserting driver through opposite bushing. New bushings are installed with chamfer toward inside. Check the special pin (20) for damage in area where it is contacted by load control shaft and renew if necessary. Also check contact areas of negative stop screw (7) and load control arm (12). Check load control shaft (21) to be sure it is not bent or otherwise damaged. Wear or damage to any other parts will be obvious.

Reassemble load sensing assembly by reversing the disassembly proce-

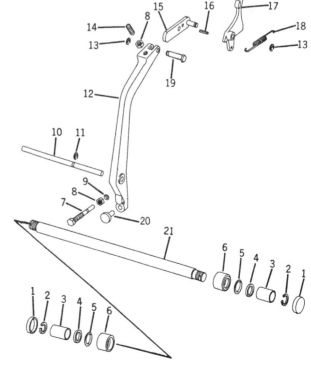

Fig. 206—Load control mechanism is located in rear of transmission case. Flexing of control shaft (21) actuates load control arm (12).

1. Plug (no rockshaft)
2. Retaining ring
3. Retaining bushing
4. Seal
5. "O" ring
6. Bushing
7. Negative stop screw
8. Jam nut
9. "O" ring
10. Pivot shaft
11. Snap ring
12. Load control arm
13. Retaining ring
14. Adjusting screw
15. Extension
16. Spring pin
17. Cam follower
18. Spring
19. Pin
20. Special pin
21. Load control shaft

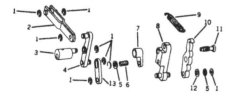

Fig. 205—Exploded view of control valve linkage.

1. Retaining ring
2. Roller (load selector) link
3. Pivot block
4. Link
5. Washer
6. Spring
7. Special nut
8. Link
9. Spring
10. Valve adjusting cam
11. Adjusting screw
12. Bowed washer
13. Link

dure and when installing hitch draft links, tighten retaining nuts until end play is removed between link collar and retaining ring, then tighten nut until next slot aligns with cotter pin hole and install cotter pin. After assembly is completed, adjust the negative stop screw as outlined in paragraph 175.

REMOTE CONTROL SYSTEM

Tractors may be equipped with single or dual selective (remote) control valves, mounted on a bracket attached to right final drive housing. Each control valve will operate a single or double acting remote cylinder and contained within the control valves are a combination flow control valve and check valve, a metering valve and two sets of operating valves. See Fig. 208 for an exploded view of the selective control valve.

SELECTIVE CONTROL VALVE TEST

All Models So Equipped

185. The selective control valve can be used to check the main hydraulic pump pressure and in addition, the operation of the selective control valve can be tested either by checking flow or by making a time cycle of the remote cylinder.

186. To check the main hydraulic pump pressure via the selective control valve, proceed as follows: Use a gage capable of registering at least 3000 psi connected to a line fitted with a male disconnect fitting. Connect line fitting to disconnect coupling, start engine and run at 2500 rpm (clutch engaged). Move control valve lever to pressurize line, hold lever in this position and check the gage reading which should be 2200-2300 psi. If main hydraulic pump pressure is not as stated, adjust pump stroke control valve as outlined in paragraph 172 or overhaul pump as outlined in paragraph 180.

187. To check flow of selective control valve, use a hydraulic tester unit and proceed as follows: Use a modified R1961R adapter and connect outlet line of tester to the large hex head plug hole directly above transmission filter as outlined in paragraph 172. Connect inlet line of tester to breakaway coupling, then start engine and run at 2500 rpm. Actuate control lever to pressurize tester, hold lever in this position, then close tester control valve until tester gage registers 2000 psi and observe tester flow gage which should read 5½-6½ gpm. If oil flow through selective control valve is not as stated, turn metering valve shown in Fig. 207 clockwise to decrease flow or counter-clockwise to increase flow.

188. If no hydraulic tester is available, oil flow through selective control can be tested by checking the extending time of a remote cylinder.

With engine running at 2100 rpm, operate selective control valve to extend remote cylinder. A 2½x8 inch remote cylinder should fully extend in approximately two seconds. If necessary, adjust the control valve metering valve shown in Fig. 207 clockwise to decrease flow or counter-clockwise to increase flow.

R&R AND OVERHAUL

Early Models

189. To remove selective control valve, or valves, disconnect breakaway

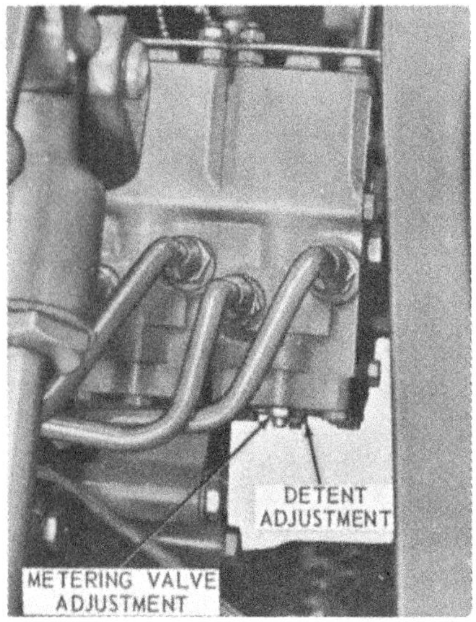

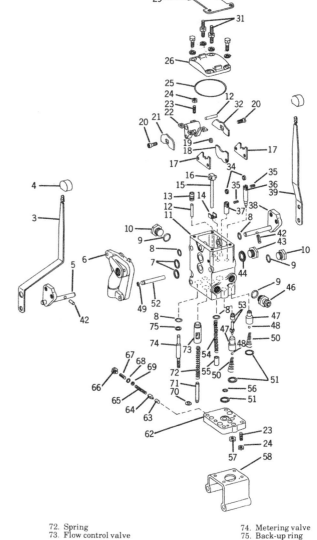

Fig. 208–Exploded view of selective (remote) control valve. Two valves can be used and are similar except arm (38) is on left side of inner valve when two valves are used.

3. Lever
4. Knob
5. Arm
6. Cover
7. "O" ring
8. "O" ring
9. "O" ring
10. Plug
11. Housing
12. Pin
13. Guide
14. Clip
15. Guide
16. Pin
17. Detent
18. Centering cam
19. Cam roller
20. Special cap screw
21. Lower valve cam
22. Rocker
23. Set screw
24. Jam nut
25. Seal
26. Cover
29. Stop
31. Special cap screw
32. Upper valve cam
34. Cam follower roller
35. Spring pin
36. Cam follower, return
37. Cam follower, pressure
38. Arm
39. Lever
42. Spring pin
43. Plug
44. Special washer
46. Connector
47. Valve
48. Ball
49. "O" ring
50. Spring
51. "O" ring
52. Shaft
53. Metering shaft & guide
54. Detent spring
55. Plug
56. "O" ring
57. Jam nut
58. Bracket
62. Cap
63. Valve seat
64. Thermal relief valve
65. Spring
66. Plug
67. Spring pin
68. "O" ring
69. Shim (.020)
70. Retaining ring
71. Spring guide
72. Spring
73. Flow control valve
74. Metering valve
75. Back-up ring

Fig. 207–View showing location of selective control valve, metering valve and detent adjustments.

coupling lines, inlet lines and return lines from valve, then unbolt valve bracket from final drive housing and lift valves and bracket from tractor.

To overhaul the selective control valve of the type shown in Fig. 208, proceed as follows: Remove control lever (3 or 39) from control arm (5 or 38), bracket (58) and side cover (6). Remove jam nut (57) from metering valve screw (74), then remove cap (62) from housing. See Fig. 209. Remove detent plug (55—Fig. 208) and spring (54), pressure and return valve springs (50), pressure and return valves (47) and the four steel balls (48). Remove square cut seals (51). Remove flow control valve spring guide (71), spring (72) and valve (73). Remove pressure and return valves metering shafts (lower part of item 53) and return valve cam followers (36). Remove metering valve (74). Remove lever stop (29), top cover (26) and seal (25), then drive out roll pin (42) and remove control arm (5 or 38). Remove plug (43), then remove cam screws (20) (one each side) which retain operating cams (21 and 32) to rocker (22) and lift out cams and rocker. Remove pressure valve cam followers (37), disconnect detent spring guide (15) from detents (17), then remove shaft (52) and the detents and centering cam (18). Remove centering pin (12), guide (13) and the control valve metering shaft guide (upper part of item 53).

Clean all parts and inspect. Also use the table given at the end of this paragraph as a guide for renewing parts. Inspect housing side cover and end caps for cracks, nicks or burrs. Small nicks or burrs can be dressed with a fine file, however if items are cracked or otherwise damaged, renew as necessary. Inspect valves (47), valve seats and balls for grooves, scoring or excessive wear and renew parts as necessary. Valve seats can be lapped if not badly

pitted or worn. Inspect metering shafts and guides (53) for wear, scoring or broken metering shaft tips. Inspect flow control valve (73) for scoring or damaged seat. If necessary, the thermal relief valve (items 63 through 69) can be removed and inspected.

Refer to the following table for spring test data:

Pressure and return valve bore
   I.D. 0.750-0.752
Pressure and return valve
   O.D. 0.748-0.749
Flow control valve bore I.D.0.750-0.752
Flow control valve O.D. ...0.748-0.749
Cam followers bore I.D. ...0.436-0.438
Cam followers O.D. 0.433-0.435
Metering valve bore
   I.D. 0.5002-0.5025
Metering valve O.D. 0.498-0.500
Valve spring
   test ....7-9 lbs. @ 1 1/16 in.
Flow control valve spring
   test ....22½-27½ lbs. @ 2⅛ in.

Selective control valves are adjustable and if disassembled, the valves must be adjusted during assembly and requires the use of John Deere special adjusting plate JDH-15, or its equivalent. To assemble and adjust the selective control valve, proceed as follows:

Install centering cam pin guide (13—Fig. 208), pin (12) and pressure valve cam followers (37). Position detents (17) and centering cam (18), install shaft (52) and "O" ring (49) then install detent spring guide (15) and be sure clip (14) is over outside edges of detents. Position operating cams (21 and 32) and be sure cams are correctly located over pressure valve cam followers. Place centering cam roller (19) and pin (12) over centering cam, place rocker in position, then secure operating cams to rocker with the special screws (20). Use new "O" ring (8), then with lever of operating arm (5 or 38) toward top of valve housing, insert operating arm through rocker and install spring pin (42) to secure arm to rocker.

Secure side cover (6) to valve body 180 degrees from normal position to serve as a holding fixture, then clamp side cover in a soft jawed vise to hold valve assembly in a position about 30 degrees from vertical with control valve end of housing upward. This will provide access to both ends of valve assembly so valve can be adjusted. Install return valve cam followers (36) so rollers are in position against operating cams, then install return valves (47) and steel balls (48). Install pressure valve metering shafts and guides (53), pressure valves (47) and steel balls (48). Install detent spring (54) and plug (55). Install control lever (3 or 39) in a reversed position so lever is approximately vertical.

Loosen jam nuts (24) and back adjusting screws (24) out away from rocker (22). If necessary, use a screwdriver to pry operating cams toward adjusting screw (special screws not loosened). Install adjusting plate (JDH-15), with valve lock screws backed out, on control valve (upper) end of valve housing. Mount a dial indicator on valve housing and position dial indicator button 3 inches upward from center line of operating arm shaft (see Fig. 210). Establish the neutral position of control lever (center of travel), zero the dial indicator and take care not to disturb dial indicator during remainder of valve adjustment. Refer to Fig. 211 to determine location of pressure and return valves, then select valve lock screws on either side of adjusting plate and tighten them to 12 inch pounds of torque. Move control lever back and forth and turn valve adjusting screws so that control lever has 0.056-0.064 travel for the pressure valve, and 0.016-0.024 travel for the return valve before lever contacts valve balls and lever travel stops. Tighten the valve adjusting screw jam nuts and be sure to recheck adjust-

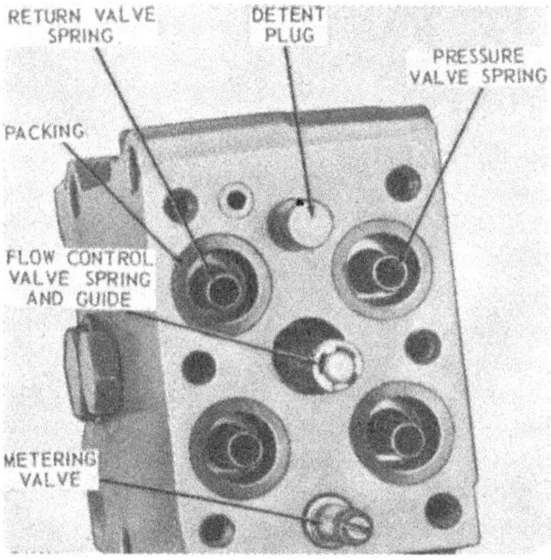

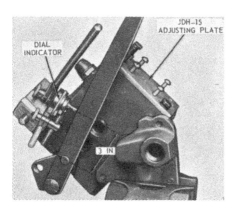

Fig. 209–Selective control valve with lower cap removed. Note square cut seals in control valve bores.

Fig. 210–Selective control valve mounted in vise showing adjusting plate and dial indicator installed. Refer to text for adjusting procedure.

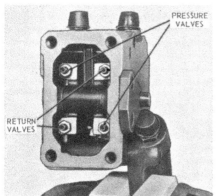

Fig. 211–View showing location of control valve pressure and return valves adjusting screws. Note relation of valves to break-away coupling ports.

ment. Tighten the two remaining valve stop screws in adjusting plate to 12 inch pounds, back-off the valve stop screws for the valves previously adjusted and adjust the remaining two valves in the same manner as described above.

NOTE: If valve adjusting screw is backed-off to increase valve clearance, the operating cam on that side must be pried against its adjusting screw without loosening the special cam screw (20—Fig. 208).

Remove the adjusting plate (JDH-15), then install the flow control valve, spring and spring guide assembly. Install the pressure and return valve springs with the tapered ends in the control valves. Be sure to include the square cut seals, install cap and bracket and tighten cap screws to 20

ft.-lbs. Install side cover and control lever in proper position, then install cover (26—Fig. 208) and seal and tighten cap screws to 35 ft. lbs. Install lever stop (29).

Reinstall selective control valve to tractor by reversing the removal procedure and adjust metering valve as outlined in paragraph 186 or 187. Refer to paragraph 190 for valve detent adjustment.

190. **DETENT ADJUSTMENT.** Refer to Fig. 207 for location of valve detent adjuster. To adjust detent spring pressure, loosen jam nut and turn detent adjusting screw in or out to hold control lever in fast extend or fast retract position. The control lever must return to neutral when remote cylinder reaches end of extending or retracting stroke. Adjustment can be made with no load on remote cylinder.

Late Models

191. Most 1020, 1520 and 2020 models after Serial Number 79542 and all 1530 and 2030 models equipped with remote control, use a selected control valve of the type shown in Fig. 212. To remove the valve or valves, disconnect breakaway coupling lines, inlet lines and return lines from valve, then unbolt valve bracket from final drive housing. Lift the valve and bracket as a unit from tractor.

To overhaul the removed valve, un-

bolt valve lever (61) and side cover (51). Remove bracket (40) and end cap (37) which will contain spring pin (45) and metering valve (50) as shown in Fig. 213. Note while loosening that the four valve guides (33—Fig. 212) are spring loaded and retained by the cap. The four guides should move out with the cap as cap screws are loosened; if they do not, protect them from flying as cap is removed, thus becoming lost or damaged. Withdraw guides, springs (30 & 32), valves (31) and associated parts, keeping them together and in proper order. Remove flow control valve (46) and spring (47). Remove snap ring (25) and detent piston outer guide (23), then withdraw detent spring (21), piston (20) and pin (19).

Invert valve body in vise, rocker assembly end up as shown in Fig. 214. Drive out the spring pin (8—Fig. 212)

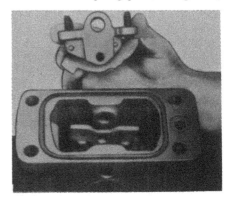

Fig. 215–Installing control rocker.

Fig. 212–Exploded view of selective (remote) control valve of the type used on late models. Two valves may be used which differ only in control lever location.

3. Cover	19. Detent pin
4. Packing	20. Detent piston
5. Operating cam	21. Spring
6. Rubber keeper	22. Special washer
7. Adjusting screw	23. Outer guide
8. Spring pin	31. Poppet valves
9. Drive pin	33. Valve guides
10. Rocker	37. End cap
11. Operating cam	40. Bracket
12. Special screw	42. Metering lever
13. Detent cam	45. Stop pin
14. Housing	46. Flow control valve
15. Spring pin	50. Metering valve
16. Roller	51. Side cover
17. Detent follower	68. Stop plate
18. Inner guide	

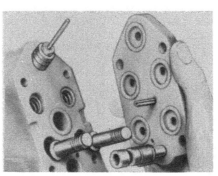

Fig. 213–Removing valve end cap.

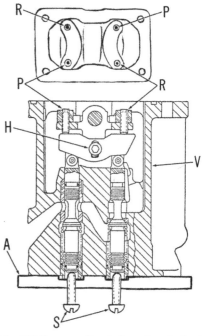

Fig. 214–View of valve rocker with top cover removed, showing adjusting screws and rubber keepers.

Fig. 216–Cross sectional view of control valve with adjusting cover installed, showing adjusting points.

A. Adjusting cover	R. Return adjusting
H. Holding screw	screws
P. Pressure adjusting	S. Seating screws
screws	V. Valve body

securing rocker (10) to lever shaft (60), withdraw the shaft and lift out rocker.

Clean all parts and inspect housing side cover and end caps for cracks, nicks or burrs. Small imperfections can be removed with a fine file, renew the parts if their condition is questionable. Inspect poppet valves (31) and their seats in housing (14) for grooves, scoring or excessive wear and renew or recondition parts as necessary.

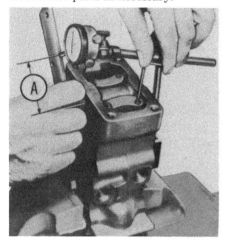

Fig. 217–Dial indicator contact point should touch control lever 3-inches from rocker axis as shown at (A).

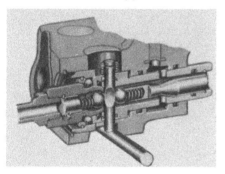

Fig. 218–Cross sectional view of breakaway coupling showing component parts. Refer to Fig. 219 for exploded view.

Selective control valve is adjustable and if disassembled, valve must be adjusted during assembly. Adjustment requires the use of a special adjusting cover (JDH-15C) or equivalent (Refer to Fig. 210), and a dial indicator as shown in Fig. 217. Install adjusting cover (A—Fig. 216) instead of valve end cap (37—Fig. 212). Snug up the retaining cap screws and tighten the four screws (S—Fig. 216) gently but firmly to hold the operating poppet valves seated. Install dial indicator with contact point touching lever 3-inches from rocker shaft as shown at (A —Fig. 217). NOTE: Lever may be reversed, if necessary for convenience in mounting dial indicator. Remove the two plugs (62—Fig. 212) and, reaching through plug holes, loosen the cam holding screws (H—Fig. 216) on each side of valve body.

Gently rock valve lever to be sure it is centered in neutral detent, then zero the dial indicator against lever arm. Carefully tighten adjusting screws (R & P—Fig. 216) equally a little at a time until cams are solid against cam followers, adjusting screws are all touching cam, and dial indicator still reads zero.

Tighten cam holding screws (H) at this time. Back the two Return Valve adjusting screws (R) out ⅛-turn and the two Pressure Valve adjusting screws (P) out ¼-turn.

With adjustment completed as outlined, rock the valve lever both ways while watching indicator dial. Indicator reading should be 0.012-0.030 inch in either direction, adjust if necessary by turning the appropriate return adjusting screw (R) a slight amount. With return adjusting screws correctly adjusted, back out the two adjusting cover screws (S) which contact return valves, then again check valve lever

movement which should now be 0.050-0.070 in either direction. Adjust if necessary, by turning the appropriate pressure adjusting screw (P).

Remove adjusting cover (A) and reassemble using new seals and gaskets. Tighten the cap screws retaining control valve cap (37—Fig. 212) to a torque of 35 ft.-lbs. and cap screws containing valve cover (3) to a torque of 20 ft.-lbs.

BREAKAWAY COUPLER

All Models

192. Fig. 218 shows a cross sectional view of coupler assembly and Fig. 219 shows an exploded view. When handle of lever (30) is crosswise to hose centerline as shown in Fig. 218, ball checks are seated, hose ends blocked and hoses can be disconnected. Also, check balls will seat if hoses are pulled from receptacle.

To disassemble the coupler, first remove unit from tractor. Punch a hole in expansion plugs (21—Fig. 219) and pry out the plugs. Remove "E" clips (22) and springs (23), then withdraw operating levers (30). Use a soft drift of the appropriate size and drive receptacle (13) rearward out of housing (24). Remove snap ring (9) and the six balls (10). Invert the receptacle (13) and insert large end of receptacle into large bore of housing, and, using housing as a holder, push down on exposed end of plug (18) and unseat and remove snap ring (19). Withdraw plug (18), spring (17) and ball (16).

"O" rings and backup rings can be renewed at this time. It is recommended that all be renewed if normal wear is the cause of failure. Assemble coupling by reversing the disassembly procedure, using Figs. 218 and 219 as a guide.

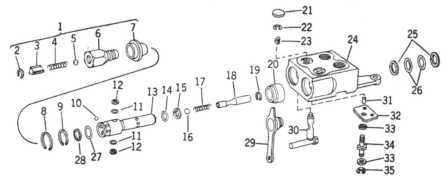

Fig. 219–Exploded view of breakaway coupler of the general type used on most models. Refer to Fig. 218 for cross section.

1. Plug assembly	9. Snap ring	16. Check ball	24. Body
2. Snap ring	10. Steel ball	17. Spring	25. Backup ring
3. Guide	11. "O" ring	18. Plug	26. "O" ring
4. Spring	12. Backup ring	19. Snap ring	27. "O" ring
5. Check ball	13. Receptacle	20. Sleeve	28. Backup ring
6. Plug	14. "O" ring	21. Expansion plug	29. Dust cover
7. Dust cover	15. Backup ring	22. "E" ring	30. Lever
8. Snap ring		23. Spring	32. Cam

REMOTE CYLINDER

All Models

193. To disassemble the remote cylinder, remove oil lines and end cap (18—Fig. 220). Remove stop valve (14) and bleed valve (13) by pushing stop rod (9) completely into cylinder. Withdraw stop valve from bleed valve being careful not to lose the small ball (12). Remove nut from piston rod and remove piston and rod. Push stop rod (9) all the way into cylinder and drift out pin (27). Remove piston rod guide (26).

Renew all seals and examine other parts for wear or damage. Wiper seal (35) should be installed with lip toward outer end of bore. Install stop rod seal assembly (1, 2 and 3) with sealing edge toward cylinder. Complete the assembly by reversing the disassembly procedure and tighten the end cap screws to 85 ft.-lbs. torque and the piston rod guide screws to 35 ft.-lbs. torque.

To adjust the working stroke, lift the piston stop lever (29), slide the adjustable stop (32) along piston rod to the desired position and press the stop lever down. If clamp does not hold securely, lift and rotate stop lever ½-turn clockwise and reset. Make certain that adjustable stop is located so that the stop rod contacts one of the flanges on adjustable stop.

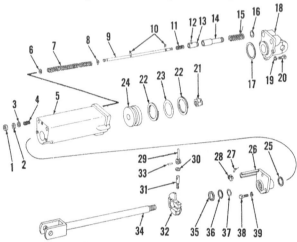

Fig. 220–Exploded view of remote cylinder used on all models.

1. Adapter	21. Nut
2. Packing	22. Back-up ring
3. Adapter	23. "O" ring
4. Spring	24. Piston
5. Cylinder	25. Gasket
6. Washer	26. Guide
7. Spring	27. Pin
8. Washer	28. Stop rod arm
9. Stop rod	29. Stop lever
10. Snap ring	30. Washer
11. Spring	31. Stop screw
12. Ball	32. Rod stop
13. Bleed valve	33. Pin
14. Stop valve	34. Piston rod
15. Spring	35. Seal
16. Gasket	36. Washer
17. Gasket	37. "O" ring
18. Cap	

NOTES

MAINTENANCE LOG

Date	Miles	Type of Service